JMP® Manual
for

Moore and McCabe's
Introduction to the Practice of Statistics
Fourth Edition

Thomas F. Devlin
Montclair State University

W. H. Freeman and Company
New York

JMP® is a registered trademark of SAS Institute Inc.

ISBN: 0-7167-9631-7

© 2003 by Thomas F. Devlin, Convent Station, New Jersey

No part of this book may be reproduced by any mechanical, photographic, or electronic process, or in the form of a phonographic recording, nor may it be stored in a retrieval system, transmitted, or otherwise copied for public or private use, without written permission from the author.

Printed in the United States of America

First printing 2002

Contents

CHAPTER 0 1

INTRODUCTION TO JMP STATISTICAL SOFTWARE 1

    0.1 Getting Acquainted with JMP and the JMP Data Table 1
    0.2 Entering and Saving Data 6
    0.3 Working with Variables and Individuals 15
    0.4 Customizing Your JMP Environment 24
    0.5 Memorizing and Rerunning Analyses 28
    0.6 Enhancing Reports and Plots 32
    0.7 Exercises 37

CHAPTER 1 39

LOOKING AT DATA—DISTRIBUTIONS 39

    1.1 Displaying Distributions with Graphs 39
    1.2 Describing Distributions with Numbers 53
    1.3 The Normal Distributions 55
    1.4 Summary 61
    1.5 Exercises 61

CHAPTER 2 63

LOOKING AT DATA—RELATIONSHIPS 63

    2.1 Displaying Relationships with Graphs 63
    2.2 Describing Relationships with Numbers: Correlation 70
    2.3 Models for Relationships: Least-Squares Regression 71
    2.4 Assessing the Fit: Residuals, Outliers, and Influential Observations 75
    2.6 Transforming Relationships 79
    2.7 Summary 82
    2.8 Exercises 82

CHAPTER 3 85

PRODUCING DATA 85

    3.1 First Steps 85
    3.2 Design of Experiments 85
    3.3 Sampling Design 88
    3.4 Toward Statistical Inference 90
    3.5 Summary 93
    3.6 Exercises 94

CHAPTER 4 95

PROBABILITY—THE STUDY OF RANDOMNESS 95

CHAPTER 5 97

SAMPLING DISTRIBUTIONS 97

5.1 Sampling Distributions for Proportions 97
5.2 The Sampling Distribution of a Sample Mean 98
5.3 Exercise 101

CHAPTER 6 103

INTRODUCTION TO INFERENCE 103

6.1 Estimating with Confidence 103
6.2 Tests of Significance 107
6.3 Summary 110
6.4 Exercises 110

CHAPTER 7 113

INFERENCE FOR DISTRIBUTIONS 113

7.1 Inference for the Mean of a Population 113
7.2 Comparing Two Means 120
7.3 Testing for Unequal Variances 124
7.4 Summary 125
7.5 Exercises 125

CHAPTER 8 127

INFERENCE FOR PROPORTIONS 127

8.1 Inference for a Population Proportion 127
8.2 Comparing Two Proportions 127
8.3 Summary 130
8.4 Exercises 130

CHAPTER 9 131

INFERENCE FOR TWO-WAY TABLES 131

9.1. Data Analysis for Two-Way Tables 131
9.2 Inference for Two-Way Tables 139
9.3 Summary 141
9.4 Exercises 141

CHAPTER 10 143

SIMPLE LINEAR REGRESSION 143

10.1 Simple Linear Regression 143
10.2 More Detail about Simple Linear Regression 150
10.3 Summary 151
10.4 Exercises 151

CHAPTER 11 153

MULTIPLE REGRESSION 153

11.1 Inference for Multiple Regression 153
11.2 A Case Study 153
11.3 Summary 162
11.4 Exercises 162

CHAPTER 12 163

ONE-WAY ANALYSIS OF VARIANCE 163

12.1 Inference for Comparing Two or More Means 163
12.2 Comparing the Means 168
12.3 Summary 174
12.4 Exercises 174

CHAPTER 13 177

TWO-WAY ANALYSIS OF VARIANCE 177

13.1 The Two-Way ANOVA Model 177
13.2 Inference for Two-Way ANOVA 177
13.3 Summary 181
13.4 Exercises 181

CHAPTER 14 183

NONPARAMETRIC TESTS 183

14.1 The Wilcoxon Rank Sum Test 183
14.2 The Wilcoxon Signed Rank Test 184
14.3 The Kruskal-Wallis Test 186
14.4 Exercises 186

CHAPTER 15 187

LOGISTIC REGRESSION 187

15.1 The Logistic Regression Model 187
15.2 Inference for Logistic Regression 187
15.3 Multiple Logistic Regression 190
15.4 Summary 191
15.5 Exercises 192

Preface

This manual is a companion to Moore and McCabe's *Introduction to the Practice of Statistics*, Fourth Edition (abbreviated IPS throughout this manual). It is intended to guide students in the use of *JMP* Software to automate the statistical graphics and analysis in the textbook.

JMP was created by John Sall in 1989 as a tool for discovering information in data through visualization and graphics. It is designed to be a point-and-click, walk-up-and-use product that enables a user to discover more, interact more, and understand more. Unlike most statistical software, *JMP* is task oriented, not method oriented.

JMP runs under both the Windows and Macintosh operating systems. The version supplied with this guide, *JMP INTRO*, is intended for introductory statistics courses. Data table creation is limited to 1000 rows. *JMP IN*, published through Duxbury Press, serves students' needs after the introductory course. The professional version of *JMP* is used extensively by industry and academic researchers. It adds specialized and advanced methods, such as industrial quality assurance, cluster analysis, and proportional hazards models.

Special thanks go to Professors Moore and McCabe for crafting an engaging introductory statistics textbook and to John Sall and the *JMP* Software team at SAS Institute for creating a software system that allows users to focus on understanding the data. I would like to thank Danielle Swearengin of W. H. Freeman for her guidance and encouragement. I would also like to thank my colleague, Andrew McDougall, and Caroline Kamowski, a student at Montclair State University, for reviewing this manual and providing valuable feedback.

The IPS CD-ROM contains text files with data from most of the examples, exercises, and tables of the textbook. In addition, the *JMP INTRO* installation CD-ROM contains *JMP* data tables and scripts for the early chapters. These are installed with the *JMP INTRO* application in folders named Sample Data and Sample Scripts, respectively. We have adopted the following naming convention throughout this manual.

Textbook Source	File Name
Example 1.6	eg01_006.txt
Exercise 1.38	ex01_038.txt
Table 1.1	ta01_001.txt
Figure 1.1	fg01_01.jmp

Chapter 0

Introduction to JMP Statistical Software

Statistics is best learned by practicing with real data. The purpose of this manual is to provide instructions on how to use JMP statistical software to automate the statistical calculations and graphics presented in *Introduction to the Practice of Statistics*, Fourth Edition, by David S. Moore and George P. McCabe. I hope that this allows you to concentrate on the meaning and purpose of the calculations.

This manual parallels the fourth edition of IPS. Chapters 1 through 15 of this manual correspond to a chapter in IPS. These chapters guide you in the use of JMP for the calculations and graphics of the corresponding chapter in IPS. This chapter serves as an introduction to JMP statistical software and discusses
- Getting acquainted with JMP and the JMP data table
- Entering and saving data
- Working with variables and individuals
- Customizing your JMP environment
- Memorizing and rerunning analyses
- Enhancing reports and graphs

If you read the first four sections thoroughly, using JMP will quickly become second nature. You will be able to produce appropriate graphs and calculations at the click of a mouse and the touch of a button. The last two sections may be read when needed.

0.1 Getting Acquainted with JMP and the JMP Data Table

0.1.1 Getting Started and Quitting

JMP is started like any other application on your operating system. Either
- Double-click a JMP data table or script, or
- Double-click the JMP INTRO icon.

To quit JMP,
- Select **File** ⇨ **Exit** or press **ctrl-Q** in Windows, or
- Select **File** ⇨ **Quit** or press **command-Q** on the Macintosh.

0.1.2 The JMP Data Table

Data to be processed in JMP must be in a *JMP data table*. A *data table* is similar to a spreadsheet but the rows and columns have a special purpose. Start JMP and open the JMP data table **Big Class** in the Sample Data folder located with the JMP INTRO application on your hard drive.

1. Select **File** ⇨ **Open** from the menu bar.

Under Windows:

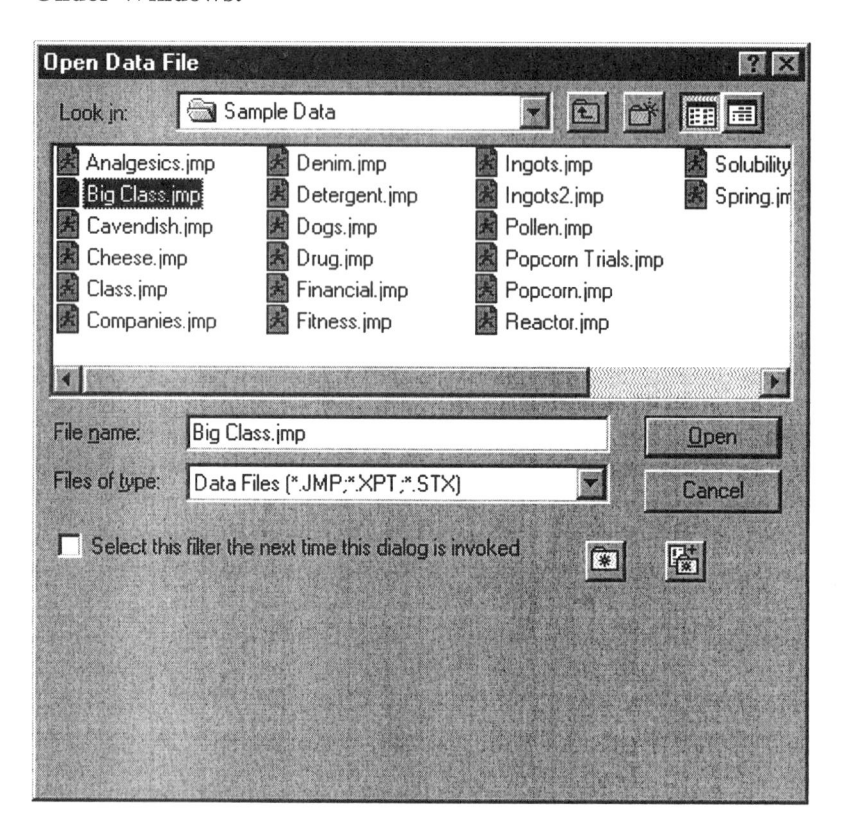

On the Macintosh:

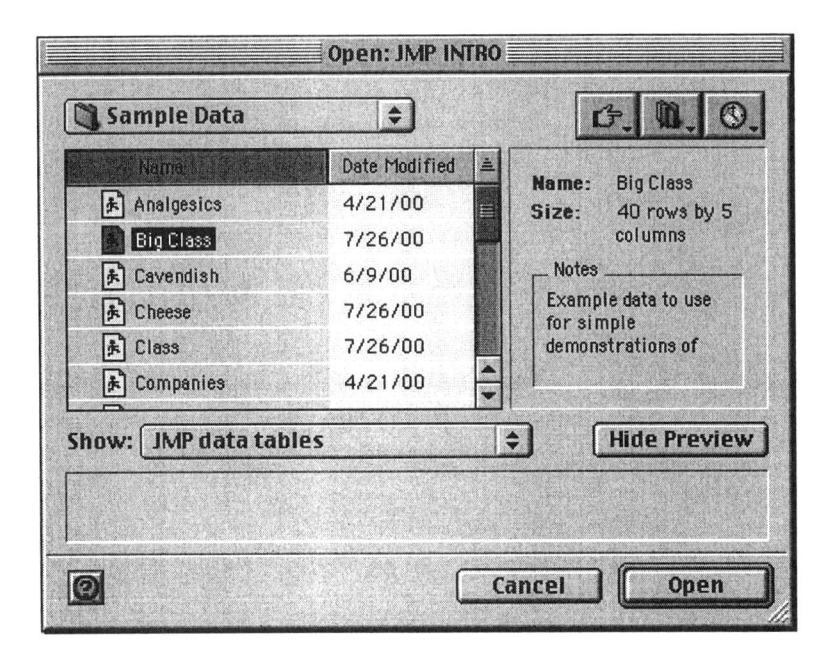

2. Select the folder **Sample Data**.

3. a. (Windows) Select **Data Files** in the `File of type` menu located below the list of files.
   b. (Macintosh) Select **JMP data tables** in the `Show` menu located below the list of files.

4. Select the file **Big Class**.

5. Press the **Open** button.

Note: Make a note of the complete path of the Sample Data folder for your computer.
_____
_____

Inspect the data table.

The *data table* looks like a spreadsheet with some enhancements. In the upper-left-hand corner, you can see that the data table has 40 rows and 5 columns. Look more closely and notice that each of the columns—**name, age, sex, height,** and **weight**—contains the values of a *variable* and each of the rows is an *individual*. Therefore, **Big Class** contains 5 variables and 40 individuals. In general, the columns of a data table contain *variables* and the rows contain *individuals*.

0.1.3 Menu Headings

JMP provides a menu bar and an icon bar of commands. The two pull-down menus at either end of the menu bar should look familiar. Let's examine the items on the menu bar.

File   performs most routine file functions, such as opening, closing, printing, and saving.

Edit   performs most editing functions, such as cutting and pasting.

Tables   perform table functions, such as sorting, subsetting, and merging.

Rows   perform row operations (i.e., operations on individuals).

Cols   perform column operations (i.e., operations on variables).

Analyze   performs most statistical analyses.

Graph   generates a variety of graphs.

Tools   display a palette of tools.

Window   selects among, organizes, and performs routine window operations on opened windows.

Help   accesses the main help feature in JMP.

Instructions in this manual will focus on accessing commands through the menus. As you become more familiar with JMP, you may wish to explore the icon alternatives.

0.1.4 Column Attributes

Inspect the data table **Big Class**. Notice that each column/variable has a *name*. Also, note that some of the columns are left-aligned and some are right-aligned. Alignment is determined by *data type*. The *data type* of a column, or variable, determines how its values are formatted in the data table, how they are stored internally, and whether they can be used in calculations. The two *data types* of interest to us are

   Numeric   for columns/variables with numeric values that can be used in calculations. These data are right-aligned. **Age, height,** and **weight** are *numeric* variables.

Character for columns/variables with numeric and/or character values that can be used to describe different levels of the variable. These data are left-aligned. **Name** and **sex** are *character* variables.

Name and *data type* are two attributes of a column/variable. To see other attributes, use the **Column Info** command found in the **Cols** menu.

Example Obtaining information about a column

Obtain column information on the variables **name** and **weight**.

1. Highlight the columns **name** and **weight** by clicking first on the top of the column **name** and then ctrl-clicking (command-clicking on a Macintosh) on the top of the column **weight.**

2. Select **Cols** ⇨ **Column Info.**

Inspect the dialog.

Notice that
- the variable **name** has *data type* character and **weight** has *data type* numeric,
- the *format* for **weight** is *Fixed Dec* with a width of 5 and no decimal places, and
- both variables have *notes* attached to them.

The *modeling type* of a variable is very important. It is not just a descriptive tag but rather it tells JMP how to analyze and graph the data. For example, the **Distribution** command displays histograms and boxplots for "Continuous" variables and stacked bar charts for "Nominal" variables.

Note: The default *modeling type* of numeric data is "Continuous" and the default *modeling type* of character data is "Nominal"

0.2 Entering and Saving Data

To process data in JMP, it must be in a *JMP data table*. You build a JMP data table either by

- **Importing data** from another application with the **Open** or the **Database** command in the **File** menu, or

- **Creating a new table** with the **New** command in the **File** menu and filling it with values by typing or pasting values into the data grid, constructing a formula, or using an external measuring instrument.

0.2.1 Importing Data

The **Open** command in the **File** menu, allows you import text files as well as open existing JMP data tables. You will be importing text files—often ones from the IPS CD-ROM.

Windows Text Import

JMP offers two choices for importing text files under MS Windows. From the **File of Type** drop-down list, you may choose

- **Text Import** opens the file and creates a JMP data table using the default rules (set in the preferences panel) to interpret end-of-field and end-of-line delimiters. This is adequate for a rectangular text file with no missing fields, a consistent delimiter, and an end-of-line delimiter.

- **Text Import Preview** allows you to modify the default field and column specifications and displays default variable names, variable data types, and data values for the first two individuals.

IPS Example 7.19 Does dietary calcium affect blood pressure?

An experiment was conducted to investigate the relationship between calcium intake and blood pressure. One group received a calcium supplement for 12 weeks while a control group received a placebo that appeared identical. The text file **ta07_004.txt** contains the seated systolic blood pressures of the subjects before and after the 12-week period. Let's import the information in that file into a JMP data table. The data can be found on the IPS CD-ROM.

1. Select **File** ⇨ **Open** from the menu bar.

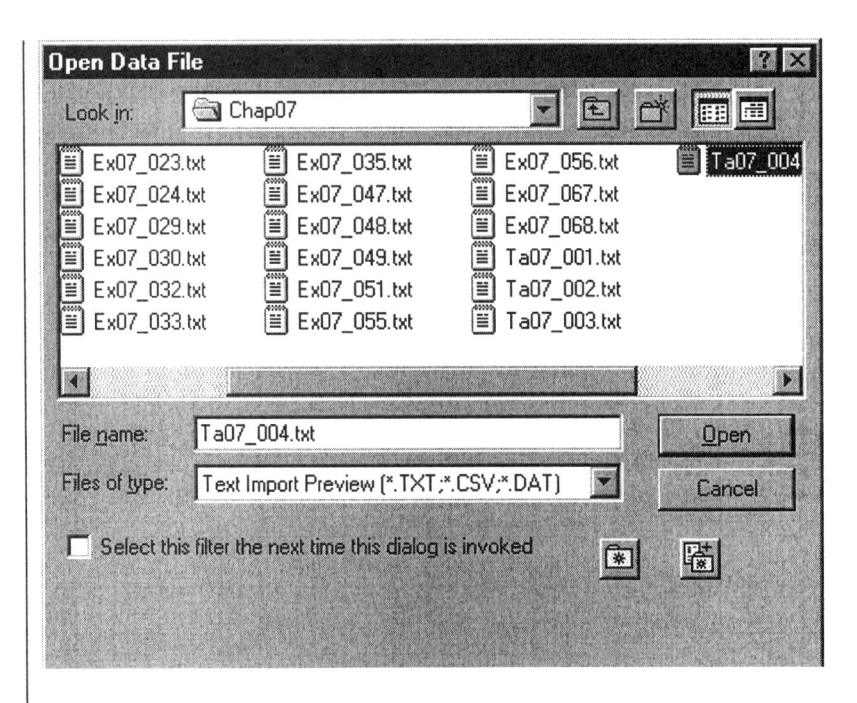

2. Select **Text Import with Preview** from the Files of Type menu.

3. Select the folder **chap07** on the IPS CD-ROM.

4. Select the file **ta07_004.txt**.

5. Click **Open**.

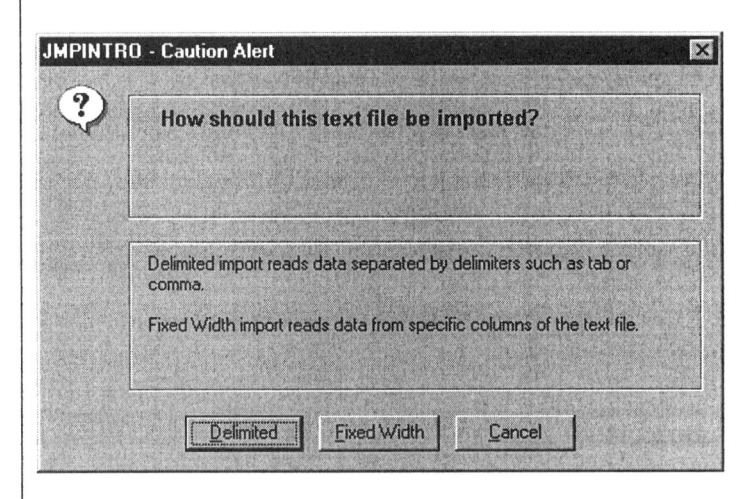

6. Click **Delimited** on the next panel.

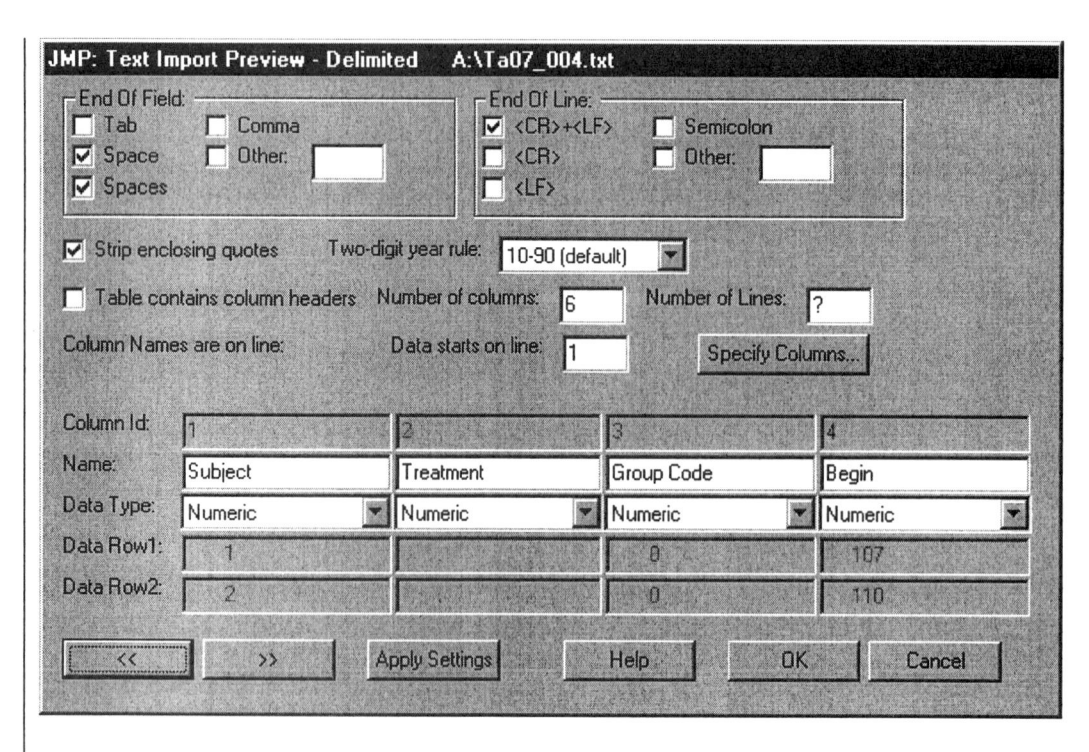

7. Change the first six column name fields to **Subject, Treatment, Group Code, Begin, End,** and **Decrease,** respectively. Press the >> button to specify the last two columns.

8. Notice that JMP has chosen data types for these columns. You may change those choices.

9. Click **OK**.

The following JMP data table is created.

Subject	Treatment	Group Code	Begin	End	Decrease
1	1	Calcium	0	107	100	7
2	0	Calcium	0	110	114	-4
3	3	Calcium	0	123	105	18
4	4	Calcium	0	129	112	17
5	5	Calcium	0	112	115	-3
6	6	Calcium	0	111	116	-5
7	7	Calcium	0	107	106	1
8	8	Calcium	0	112	102	10
9	9	Calcium	0	136	125	11
10	10	Calcium	0	102	104	-2
11	11	Placebo	1	123	124	-1
12	12	Placebo	1	109	97	12
13	13	Placebo	1	112	113	-1
14	14	Placebo	1	102	105	-3
15	15	Placebo	1	98	95	3

Notice that the data for the second group of subjects is placed directly below the data for the first group rather than adjacent to it as in Table 7.4 of the textbook. Recall that, in a JMP data table, the rows are the *individuals* and the columns are the *variables*.

Macintosh Text Import

To import text data on the Macintosh, first select **Text Documents** from the **Show** menu. The **Open As** menu then appears and gives three choices:

- **Text** opens the file in a text editing window without creating a JMP data table,

- **Text Data** opens the file and creates a JMP data table using the default rules (set in the preferences panel) to interpret end-of-field and end-of-line delimiters, and

- **Text Data with Preview** opens the file, creates a JMP data table using delimiters that you designate, and displays default variable names and data types and the data values of the first two individuals.

We illustrate the last choice.

IPS Example 7.19 Does dietary calcium affect blood pressure?

An experiment was conducted to investigate the relationship between calcium intake and blood pressure. One group received a calcium supplement for 12 weeks while a control group received a placebo that appeared identical. The text file **ta07_004.txt** contains the seated systolic blood pressures of the subjects before and after the 12-week period. Let's import the information in that file into a JMP data table. The data can be found on the IPS CD-ROM.

1. Select **File** ⇨ **Open** from the menu bar.

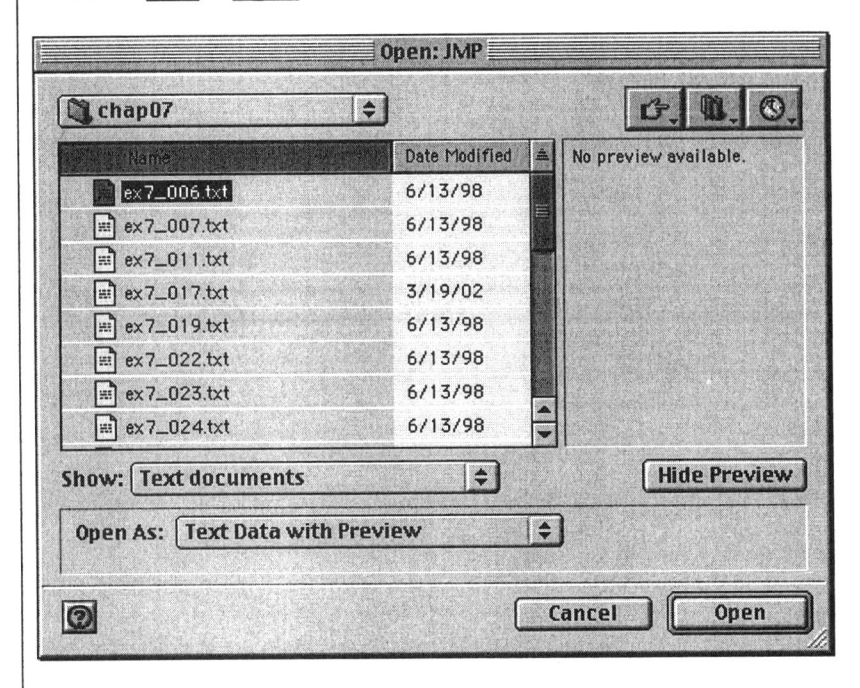

10   Introduction to JMP Statistical Software

2. Select **Text documents** from the **Show** menu.

3. Select **Text Data with Preview** from the **Open As** menu.

4. Select the folder **chap07** on the IPS CD-ROM.

5. Select the file **ta07_004.txt**.

6. Click **Open**.

7. Click **Delimited** on the next panel.

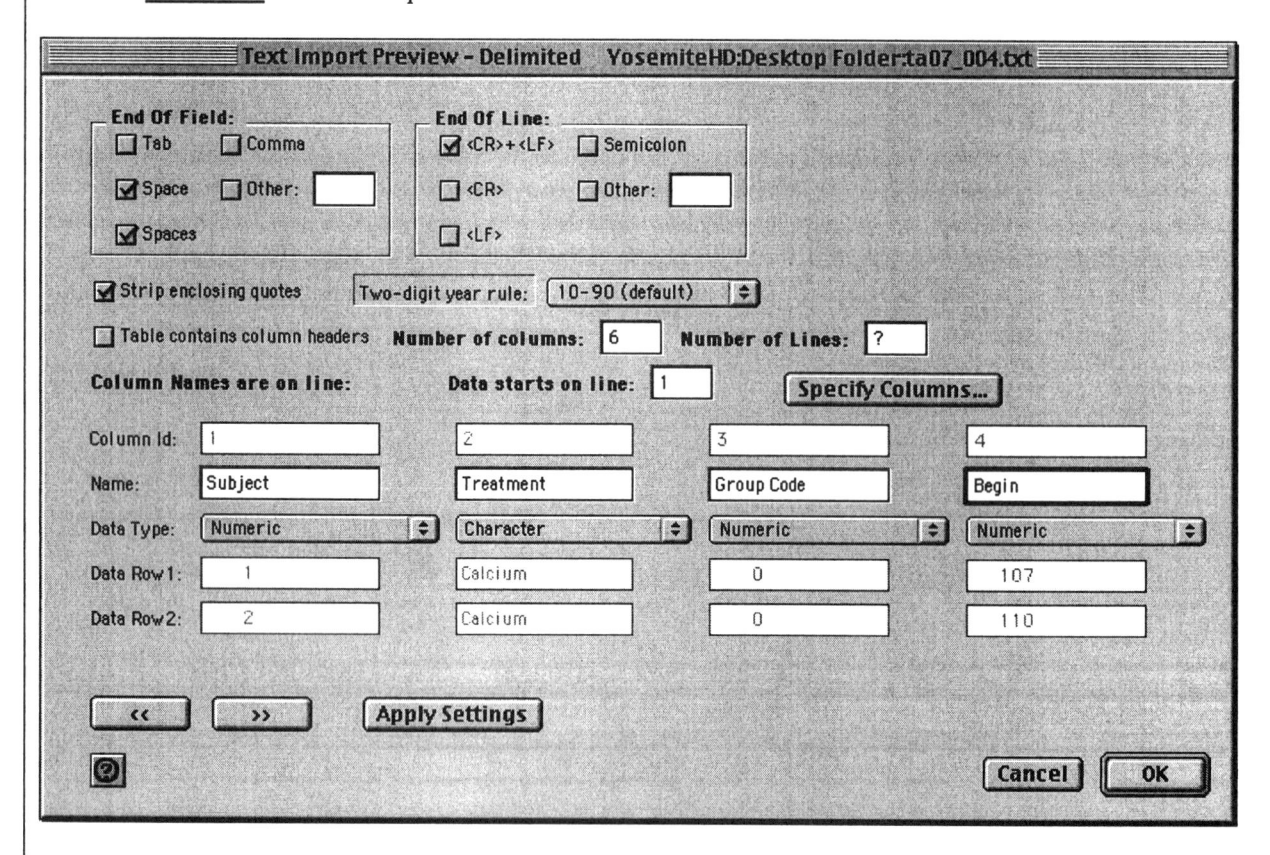

8. Change the first six column mame fields to `Subject`, `Treatment`, `Group Code`, `Begin`, `End`, and `Decrease`, respectively. Press the >> button to specify the last two columns.

9. Notice that JMP has chosen data types for these columns. You may change those choices.

10. Click **OK**.

The following JMP data table is created.

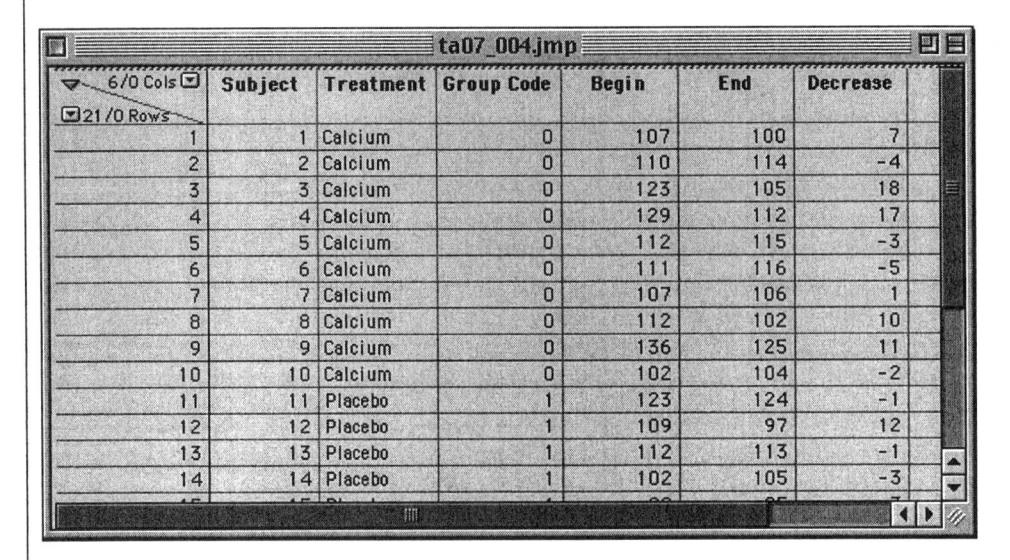

	Subject	Treatment	Group Code	Begin	End	Decrease
1	1	Calcium	0	107	100	7
2	2	Calcium	0	110	114	-4
3	3	Calcium	0	123	105	18
4	4	Calcium	0	129	112	17
5	5	Calcium	0	112	115	-3
6	6	Calcium	0	111	116	-5
7	7	Calcium	0	107	106	1
8	8	Calcium	0	112	102	10
9	9	Calcium	0	136	125	11
10	10	Calcium	0	102	104	-2
11	11	Placebo	1	123	124	-1
12	12	Placebo	1	109	97	12
13	13	Placebo	1	112	113	-1
14	14	Placebo	1	102	105	-3

Notice that the data for the second group of subjects is placed directly below the data for the first group rather than adjacent to it as in Table 7.4 of the textbook. Recall that, in a JMP data table, the rows are the *individuals* and the columns are the *variables*.

Remarks

- JMP automatically detected that **Treatment** should be given a data type of character.

- Default choices for End-of-Field and End-of-Line *delimiters* and other import settings can be changed in the preferences panel. See Section 0.4.3 for details.

0.2.2 Creating a New JMP Data Table

The **New** command in the **File** menu displays an empty data table with no rows and one column, named **Column 1**.

To add variables,
 • Use the **New Column** or **Add Multiple Columns** command in the **Cols** menu.

To add individuals,
 • Use the **Add Rows** command from the **Rows** menu or
 • Simply type in a cell anywhere beyond the last row of the table.

You can use the usual editing commands, such as cut and paste, to enter data values. You can also use drag and drop to copy or rearrange columns.

IPS Exercise 2.92 Moore's law

Gordon Moore, one of the founders of Intel Corporation, predicted that the number of transistors on an integrated circuit chip would double every 18 months. This has become known as "Moore's law" for microprocessing power. Here are data on the dates and number of transistors for Intel microprocessors.

Processor	Date	Transistors	Processor	Date	Transistors
4004	1971	2,250	486 DX	1989	1,180,000
8008	1972	2,500	Pentium	1993	3,100,000
8080	1974	5,000	Pentium II	1997	7,500,000
8086	1978	29,000	Pentium III	1999	24,000,000
286	1982	120,000	Pentium 4	2000	42,000,000
386	1985	275,000			

In Exercises 92 and 94 of Chapter 2, you will be asked to verify Moore's prediction. Let's create a JMP data table for the data.

1. Select **File** ⇨ **New** from the menu bar.

Adding Columns

2. Select **Cols** ⇨ **Add Multiple Columns** to accommodate the 3 variables **Processor, Date,** and **Transistors**.

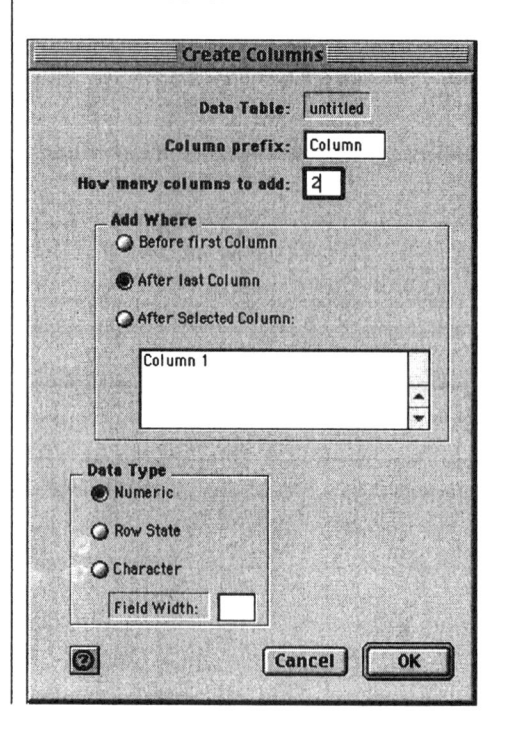

3. Enter <u>3</u> after **How many columns to add** and press <u>OK</u>.

Now let's change the name of the first column to **Processor**.

4. Select the first column of the data grid to highlight that column.

5. Click on the name **Column 1** to highlight the column name.

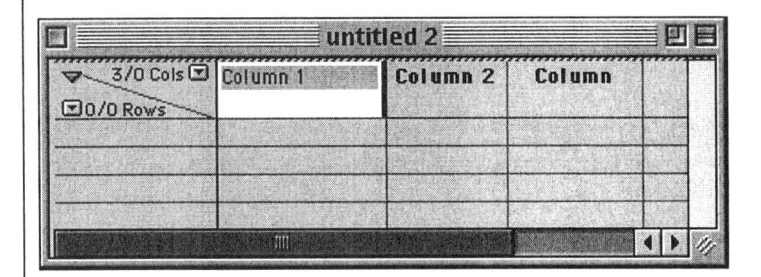

6. Type `Processor`.

7. Repeat steps 4 to 6 on the other columns to change the names to **Date** and **Transistors**.

Setting the Data Type of a Column

By default, columns contain numeric data. However, the names of the processors are character values. Use the **Column Info** command to change the data type of the variable **Processor**.

8. Select the first column.

9. Select <u>Cols</u> ⇨ <u>Column Info</u>.

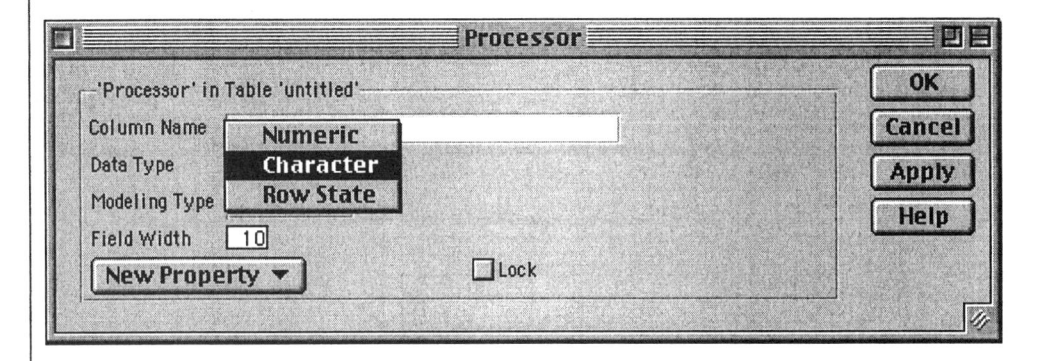

10. Select <u>Character</u> from the pop-up menu for `Data Type` and press <u>OK</u>.

Adding Rows

Adding rows is easy.

11. Select <u>Rows</u> ⇨ <u>Add Rows</u> from the menu bar and enter <u>11</u> and press <u>OK</u>.

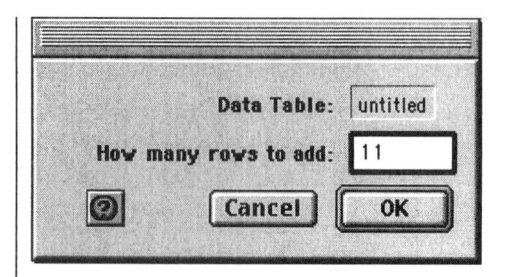

Entering Data

Entering data into the data table is similar to entering data into a spreadsheet.

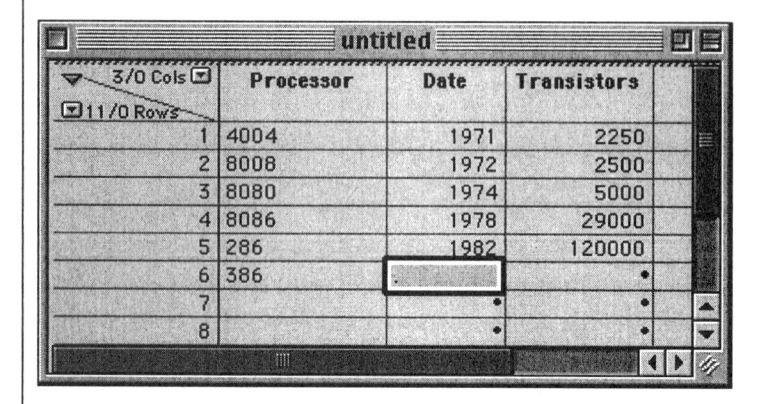

12. Select the first cell in the first row and enter **4004**.

13. Press the **Tab** key, enter **1971**, press **Tab** again, and enter **2250**.

14. Continue until you have finished entering the data.

In the next section, you will learn how to save the data table for later use.

0.2.3 Saving and Naming a Data Table

The **Save** command in the **File** menu writes the current JMP data table to a file.

Note: JMP analysis windows are not saved with the data table. However, you can use JMP tools to copy reports to other applications or you can save the JSL script that produced the analysis. These features are discussed in Sections 0.5 and 0.6.

IPS Exercise 2.92 Moore's Law (cont'd.)

To save the data table for analysis in conjunction with Chapter 2,

1. Select **File** ⇨ **Save**

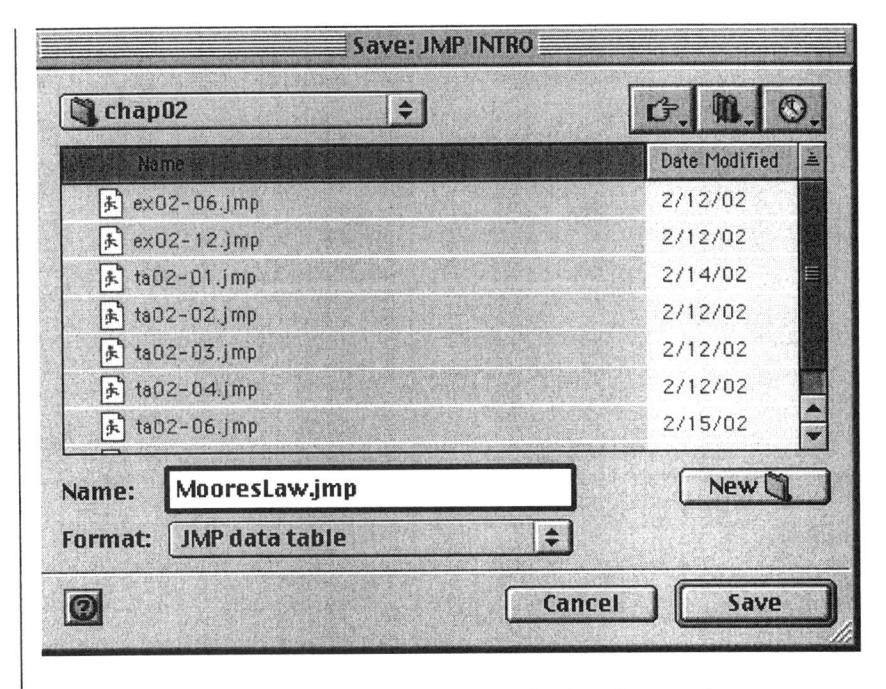

2. Type **MooresLaw.jmp** in the **Name** field and press **Save**.

0.3 Working with Variables and Individuals

0.3.1 Selecting Individuals and Columns

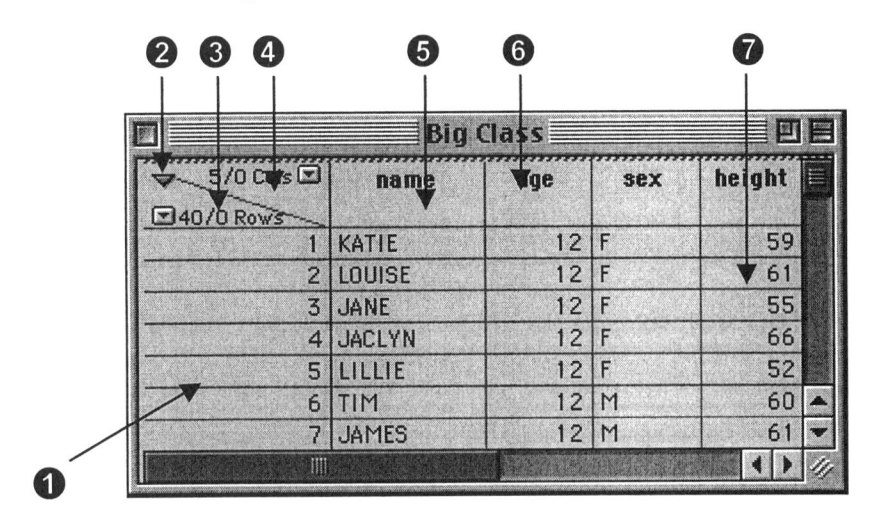

Select the area marked

❶ to select a row/individual
❷ to open the left side panels
❸ to deselect all rows/individuals
❹ to deselect all columns/variables
❺ to select a column/variable

❻ to change the column/variable name
❼ to edit a cell

After selecting an individual (or a variable), press the
- Shift key to select a block of adjacent individuals (or variables), and
- Ctrl (command on the Macintosh) key to select nonadjacent individuals (or variables).

You can also use the **Row Selection** command in the **Rows** menu and the **Select Where...** option to select rows/individuals that meet a criterion. Suppose that you wish to select the female students in the JMP data table **Big Class** that is stored in the Sample Data folder located with the JMP INTRO application.

1. Select **File** ⇨ **Open** ⇨ **Big Class**.

2. Select **Rows** ⇨ **Row Selection** ⇨ **Select Where...**

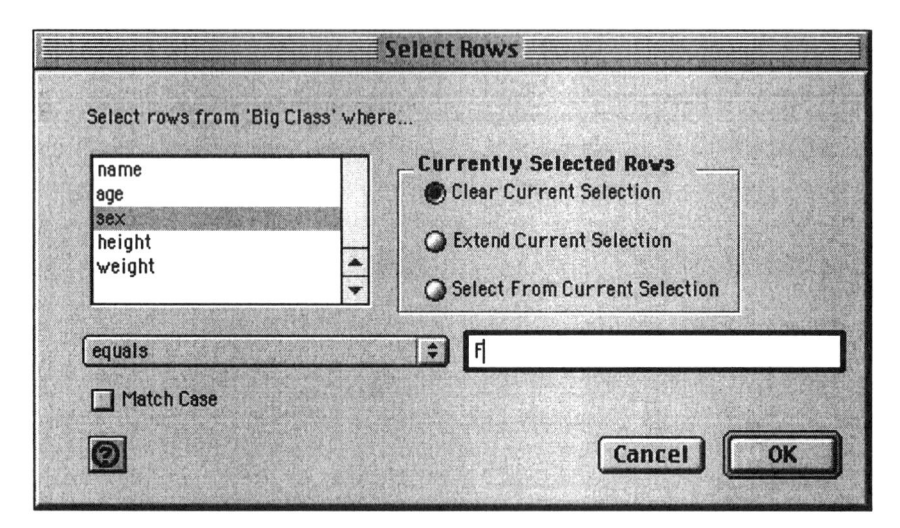

   a. Select **sex** in the list of columns.
   b. Select **equals** (the default) in the comparison menu.
   c. Type **F** as the value.
   d. Select **OK**.

	name	age	sex	height	weight
1	KATIE	12	F	59	95
2	LOUISE	12	F	61	123
3	JANE	12	F	55	74
4	JACLYN	12	F	66	145
5	LILLIE	12	F	52	64
6	TIM	12	M	60	84
7	JAMES	12	M	61	128
8	ROBERT	12	M	51	79
9	BARBARA	13	F	60	112
10	ALICE	13	F	61	107
11	SUSAN	13	F	56	67
12	JOHN	13	M	65	98
13	JOE	13	M	63	105
14	MICHAEL	13	M	58	95
15	DAVID	13	M	59	79
16	JUDY	14	F	61	81
17	ELIZABETH	14	F	62	91
18	LESLIE	14	F	65	142
19	CAROL	14	F	63	84
20	PATTY	14	F	62	85
21	FREDRICK	14	M	63	93
22	ALFRED	14	M	64	99

0.3.2 Changing the State of an Individual

There are times when we wish to
- *exclude* one or more individuals from analysis,
- *hide* one or more individuals in a plot, or
- *color* points representing one or more individuals in a plot.

These tasks are easily accomplished in JMP by changing the *state* of an individual. The operations involve individuals so commands to perform them are found on the **Rows** menu.

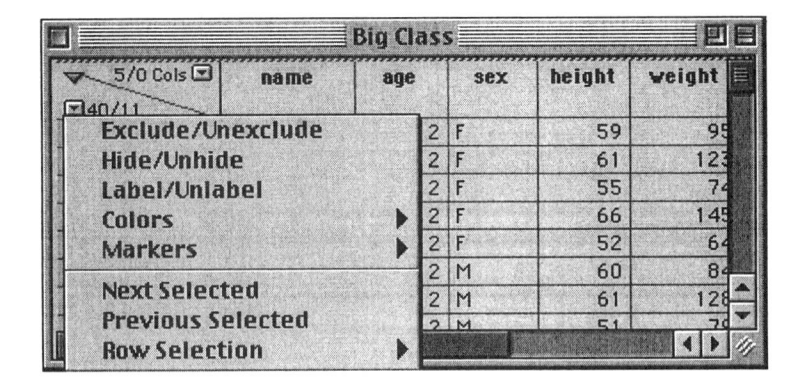

To illustrate these features, we will first produce an analysis with all individuals in their default states.

1. Deselect all individuals in **Big Class** by selecting <u>Rows</u> ⇨ <u>**Clear Row States**</u>.

2. Select <u>Analyze</u> ⇨ <u>Fit Y by X</u>.

3. Select the column **weight** and click **Y, Response**.

4. Select the column **height** and click **X, Factor**.

5. Press **OK**.

6. Click on the red triangle next to **Bivariate Fit of ...** and select **Fit Line** from the menu that opens.

The following report is produced.

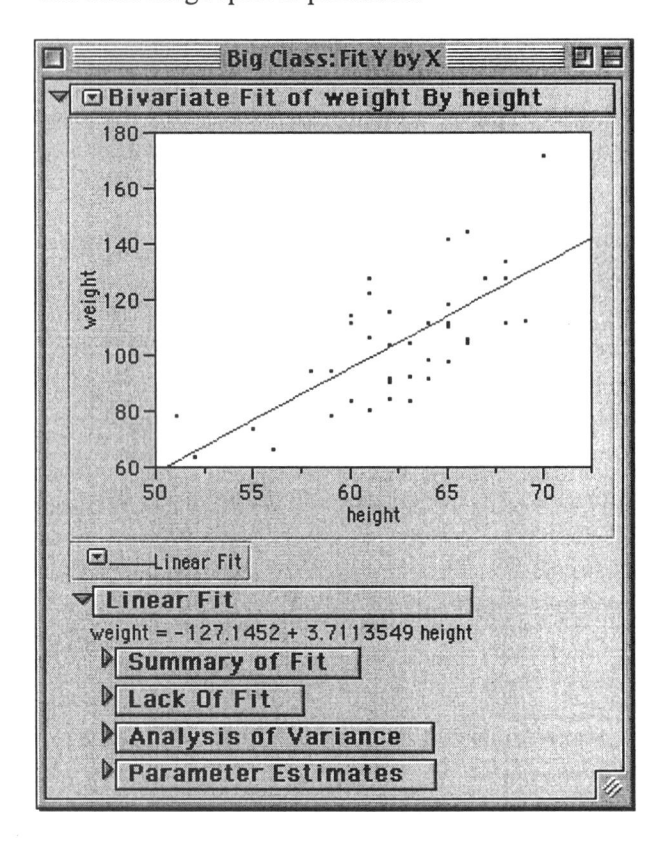

[Big Class: Fit Y by X — Bivariate Fit of weight By height scatterplot with Linear Fit: weight = -127.1452 + 3.7113549 height; includes Summary of Fit, Lack Of Fit, Analysis of Variance, Parameter Estimates sections.]

Now change the state of some individuals. Let's color the female students red and change their plotting symbol.

7. Use the **Rows Selection** command in the **Rows** menu to select the female students as we did above.

8. Select **Rows** ⇨ **Colors** ⇨ **red**; then select **Rows** ⇨ **Markers** ⇨ ✖.

Let's hide all 12-year-olds in the scatterplot.

9. Select rows where the age is 12 by using the **Rows Selection** command again.
   a. Select **Rows** ⇨ **Row Selection** ⇨ **Select Where...**
   b. Select **age** in the list of columns.
   c. Type **12** as the value.
   d. Select **OK**.

10. Select **Rows** ⇨ **Hide/Unhide**.

Finally, we will set individual 40, Lawrence, to be excluded from future analyses. Lawrence is the tallest and heaviest student.

11. Select row **40** and then **Rows** ⇨ **Exclude/Unexclude**.

The data grid updates to indicate all of the new row states.

		name	age	sex	height	weight
✕ ⊕	1	KATIE	12	F	59	95
✕ ⊕	2	LOUISE	12	F	61	123
✕ ⊕	3	JANE	12	F	55	74
✕ ⊕	4	JACLYN	12	F	66	145
✕ ⊕	5	LILLIE	12	F	52	64
⊕	6	TIM	12	M	60	84
⊕	7	JAMES	12	M	61	128
⊕	8	ROBERT	12	M	51	79
✕	9	BARBARA	13	F	60	112
✕	10	ALICE	13	F	61	107
✕	11	SUSAN	13	F	56	67
	12	JOHN	13	M	65	98
	13	JOE	13	M	63	105
	14	MICHAEL	13	M	58	95
	15	DAVID	13	M	59	79
✕	16	JUDY	14	F	61	81
✕	17	ELIZABETH	14	F	62	91
✕	18	LESLIE	14	F	65	142
✕	19	CAROL	14	F	63	84
✕	20	PATTY	14	F	62	85
	21	FREDRICK	14	M	63	93
	22	ALFRED	14	M	64	99
	23	HENRY	14	M	65	119
	24	LEWIS	14	M	64	92
	25	EDWARD	14	M	68	112
	26	CHRIS	14	M	64	99
	27	JEFFERY	14	M	69	113
✕	28	MARY	15	F	62	92
✕	29	AMY	15	F	64	112
	30	ROBERT	15	M	67	128
	31	WILLIAM	15	M	65	111
	32	CLAY	15	M	66	105
	33	MARK	15	M	62	104
	34	DANNY	15	M	66	106
✕	35	MARTHA	16	F	65	112
✕	36	MARIAN	16	F	60	115
	37	PHILLIP	16	M	68	128
✕	38	LINDA	17	F	62	116
	39	KIRK	17	M	68	134
⊘	40	LAWRENCE	17	M	70	172

Move the **Fit Y by X** window to the front so that you can see the effect of these changes on the analyses.

12. Select **Window** ⇨ **Big Class: Fit Y by X**.

Notice that the female data is red, the 12-year-olds are invisible, and all 40 individuals (rows) are still used in the analyses.

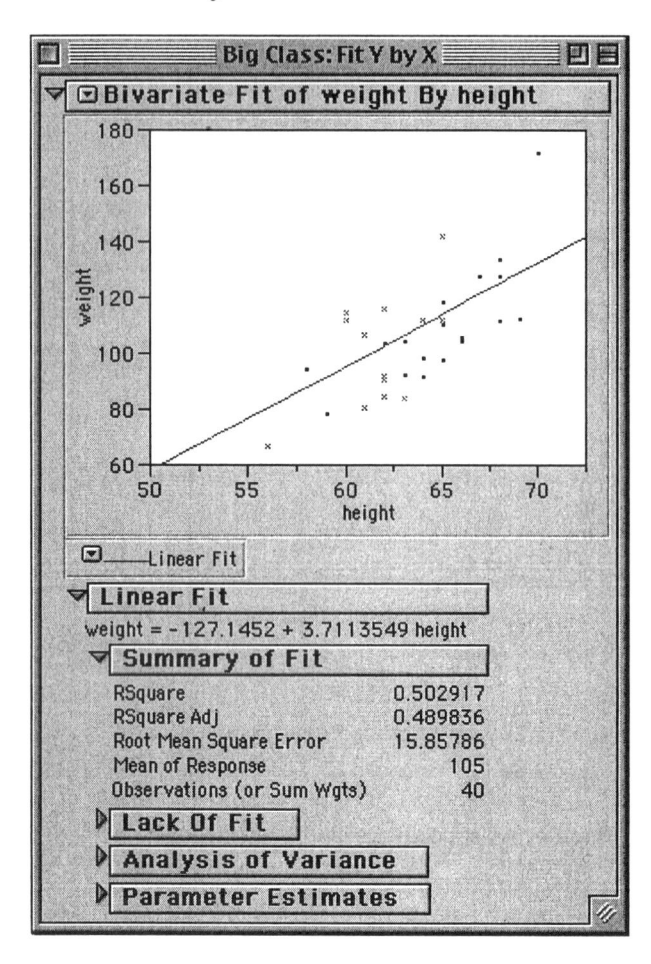

Move the cursor to the point in the upper-right-hand corner. Notice that Lawrence is still there. To see the effects of changing the state of Lawrence to exclude, we must redo the analysis with Lawrence in that state. Let's use a shortcut to do that.

13. Click on the red triangle next to **Bivariate Fit....**

14. Select **Script** ⇨ **Redo Analysis** on the menu that opens.

Notice that Lawrence is no longer in the scatterplot; the equation of the line is quite different; and only 39 observations are used in the fit.

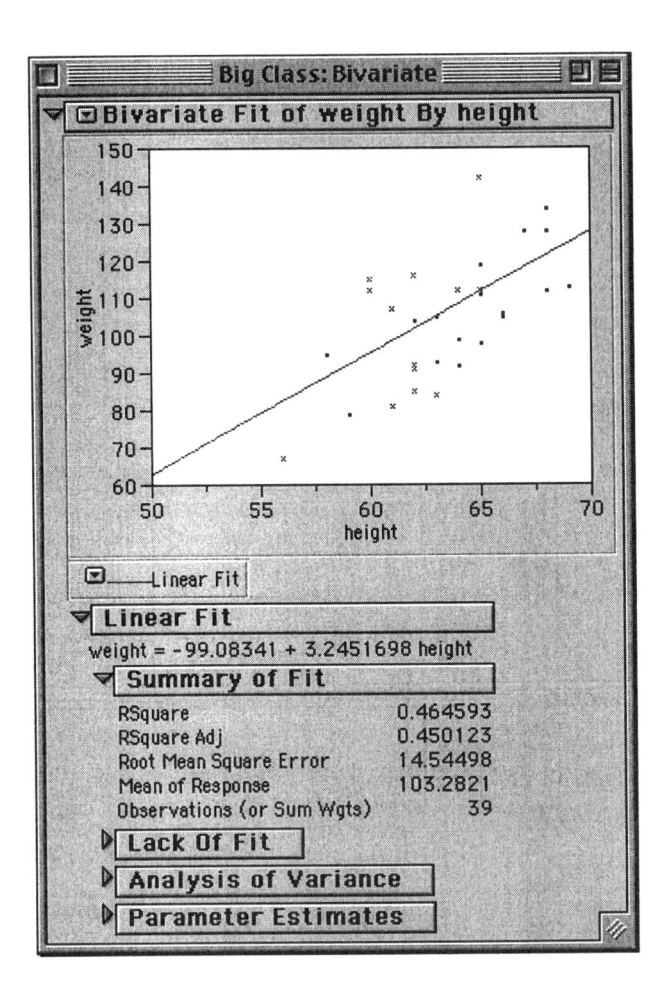

```
┌─────────────────────────────────────┐
│  □    Big Class: Bivariate    □ ▣  │
│ ▽ □ Bivariate Fit of weight By height│
│     150-                             │
│     140-                         ×  │
│     130-                      ·  ·  │
│     120-          ×        ·      ·│
│  weight 110-    ×  ×    × ·    ·   │
│     100-         ×    ×  ·  ·      │
│      90-            × ×            │
│      80-       ×  × ×              │
│      70-                            │
│      60-  ×                         │
│         50    55    60    65    70  │
│                   height             │
│  □——Linear Fit                       │
│ ▽ Linear Fit                         │
│  weight = -99.08341 + 3.2451698 height│
│ ▽ Summary of Fit                     │
│   RSquare                  0.464593  │
│   RSquare Adj              0.450123  │
│   Root Mean Square Error  14.54498   │
│   Mean of Response        103.2821   │
│   Observations (or Sum Wgts)    39   │
│ ▷ Lack Of Fit                        │
│ ▷ Analysis of Variance               │
│ ▷ Parameter Estimates                │
└─────────────────────────────────────┘
```

0.3.3 Creating a New Variable Using a Formula

Sometimes we may wish to add or subtract the values of several variables for each individual. For example, we may want to look at the differences of before and after values in a study. Often, a statistician needs to reexpress a variable. For example, the square root of the variable amount might follow a more recognizable pattern than the amounts themselves. Or the relationship of the logarithm of the dose of a new medication and the clinical response may be simpler to understand and describe than the dose-response relationship. At other times, we may wish to randomly generate data from a particular distribution. In each case, we need to construct a new variable from one or more existing variables or from a mathematical or statistical function. The *Formula Editor* in JMP is a powerful and easy-to-use tool for doing these tasks and more. We illustrate its use with a simple example.

Example Finding the ratio of the weight to height of a child

The JMP data table **Big Class** contains the heights and weights of 40 teens and preteens. It is likely that their heights and weights are related. We might wonder then if the ratio of a child's weight to his or her height is relatively constant. To investigate this, we decide to construct a new variable—**weight** divided by **height**.

1. Select **File** ⇨ **Open** and select **Big Class.jmp** from the Sample Data folder that comes with the JMP INTRO application.

2. Select **Cols** ⇨ **New Column**.

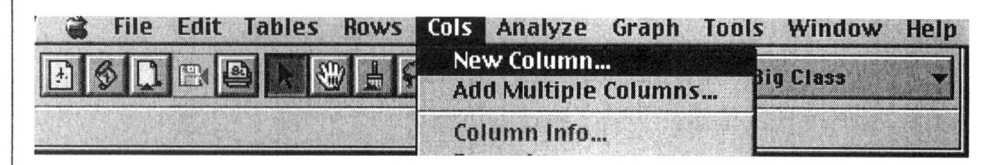

3. Name the column `Adjusted Wt` and select a format of **Fixed Dec** with **2** decimal places.

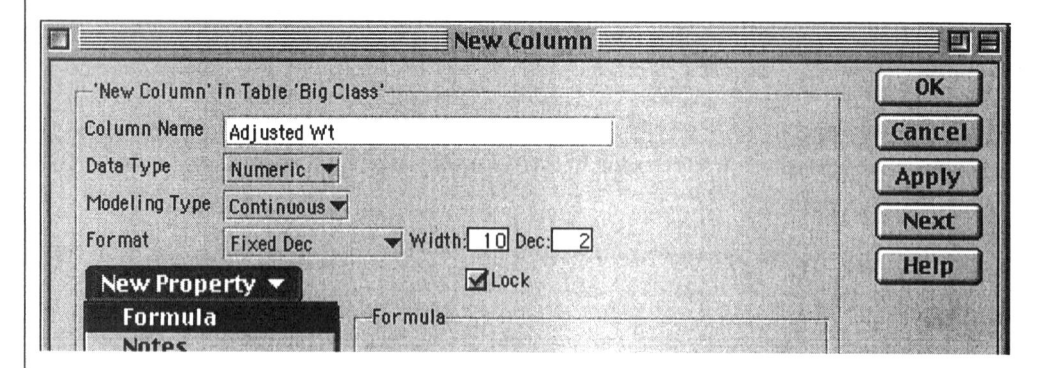

4. Select **New Property** ⇨ **Formula** to open the *Formula Editor* window.

We now build the formula to calculate the variable **Adjusted Wt**.

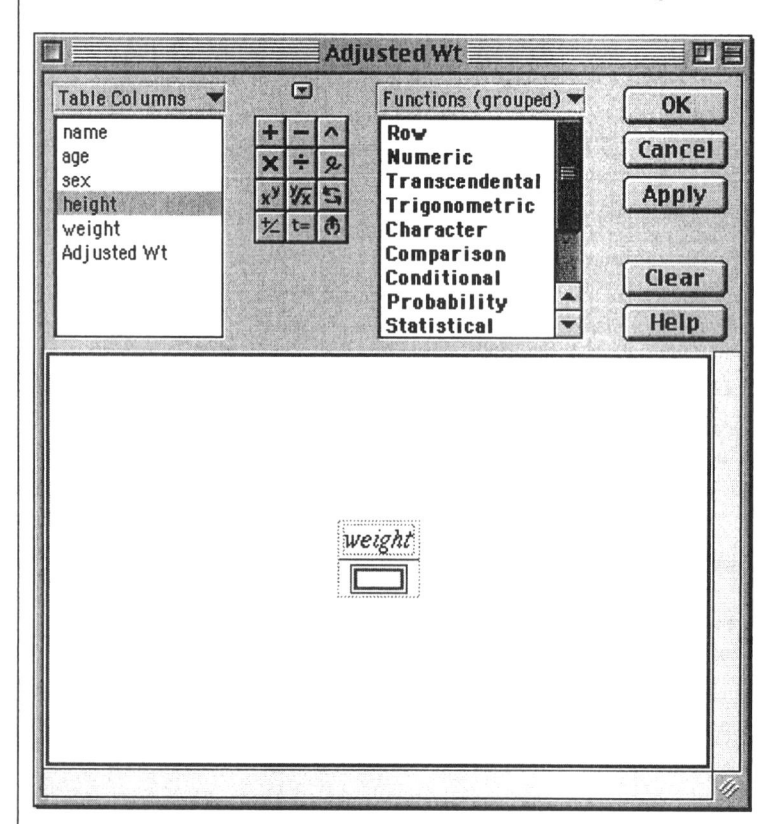

5. Select **weight** from the list of columns.

6. Press ÷ on the keypad and select **height** from the list of columns.

7. Press **Apply** and look at the data table.

The values of the ratio **Adjusted Wt** range from about 1.20 to 2.50 lbs/inch.

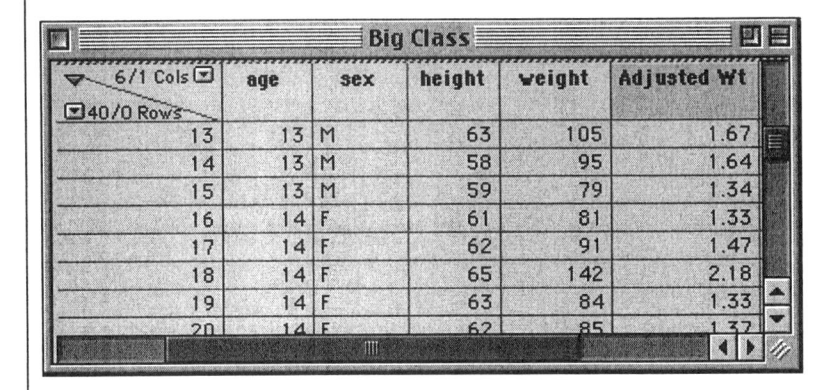

	age	sex	height	weight	Adjusted Wt
13	13	M	63	105	1.67
14	13	M	58	95	1.64
15	13	M	59	79	1.34
16	14	F	61	81	1.33
17	14	F	62	91	1.47
18	14	F	65	142	2.18
19	14	F	63	84	1.33
20	14	F	62	85	1.37

Remark

Look at the rich list of functions. Some are used in later chapters.

1. Select **Transcendental** from the list.

2. Scroll down the list and select **Random**.

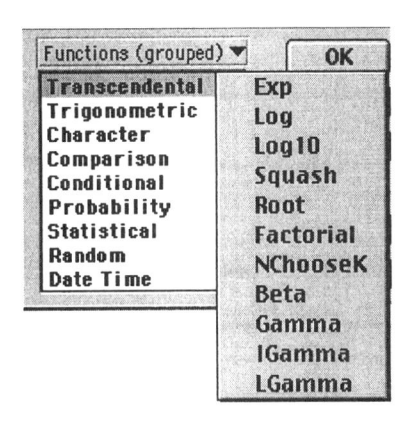

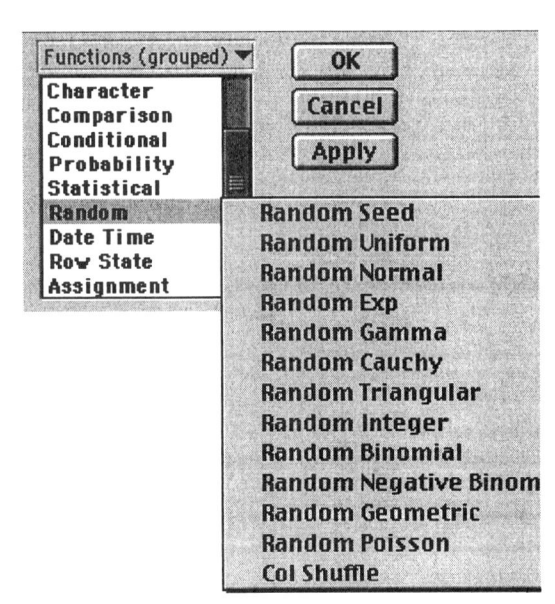

Transcendental Functions

Random Functions

0.4 Customizing Your JMP Environment

This section discusses the JMP Starter window, table information panels, and setting preferences. We make specific recommendations for and show you how to customize your session environment.

0.4.1 The JMP Starter Window

If a JMP data table is not selected before starting, JMP begins by opening a special navigation window—the *JMP Starter*.

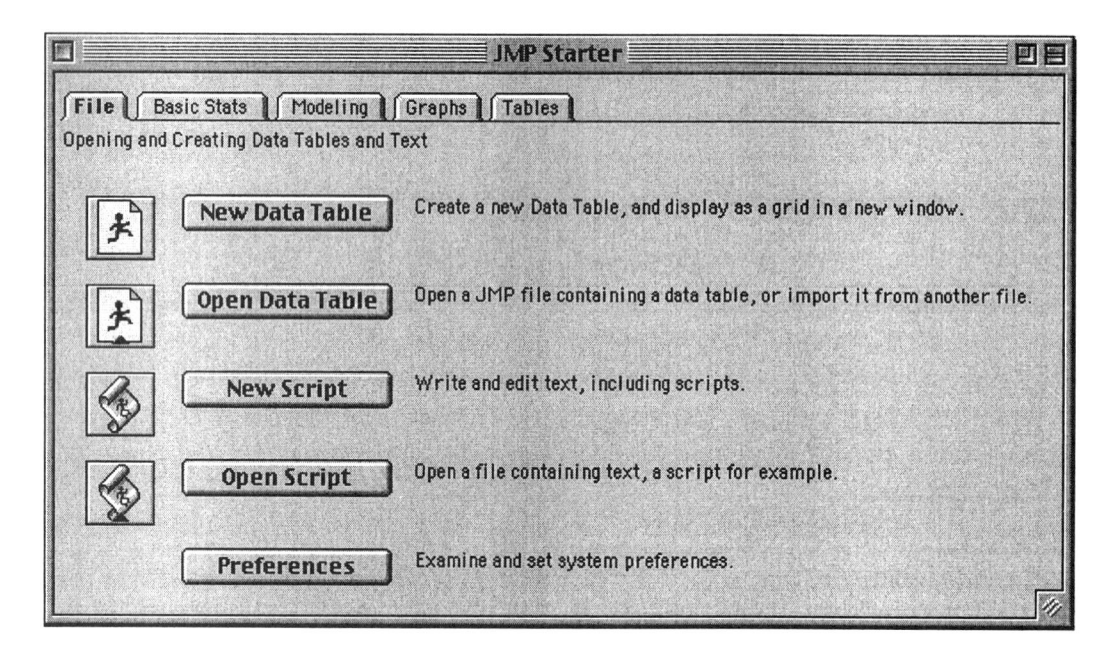

The JMP Starter window presents an alternate way to access JMP commands that we will not use. All these commands are accessible through the menu bar. We recommend that the Starter window be closed at startup. See "Setting Preferences" in Section 0.4.3 for details.

0.4.2 Table Information Panels

Table information panels on the left side of the data grid offer access to meta-data and duplicate the **Tables**, **Rows**, and **Cols** menus.

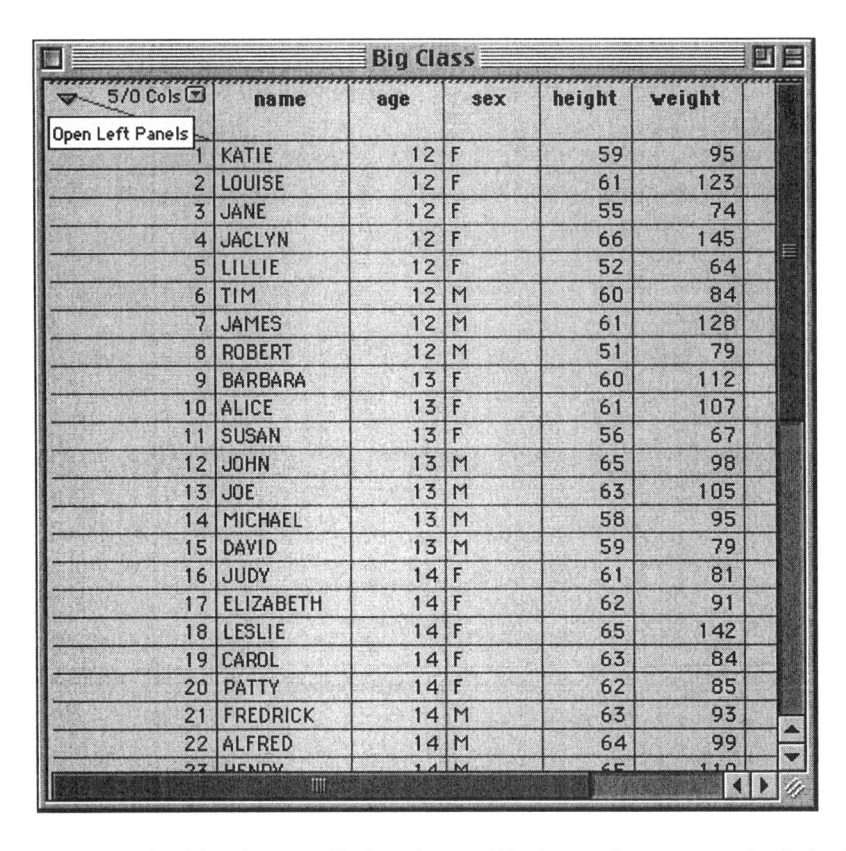

	name	age	sex	height	weight
1	KATIE	12	F	59	95
2	LOUISE	12	F	61	123
3	JANE	12	F	55	74
4	JACLYN	12	F	66	145
5	LILLIE	12	F	52	64
6	TIM	12	M	60	84
7	JAMES	12	M	61	128
8	ROBERT	12	M	51	79
9	BARBARA	13	F	60	112
10	ALICE	13	F	61	107
11	SUSAN	13	F	56	67
12	JOHN	13	M	65	98
13	JOE	13	M	63	105
14	MICHAEL	13	M	58	95
15	DAVID	13	M	59	79
16	JUDY	14	F	61	81
17	ELIZABETH	14	F	62	91
18	LESLIE	14	F	65	142
19	CAROL	14	F	63	84
20	PATTY	14	F	62	85
21	FREDRICK	14	M	63	93
22	ALFRED	14	M	64	99

Press the blue icon, called a closure/disclosure button, on the left side of the data grid portion of the data table to open and close the left side panels at any time. When you hover the mouse over the button, JMP indicates the button's function as shown above.

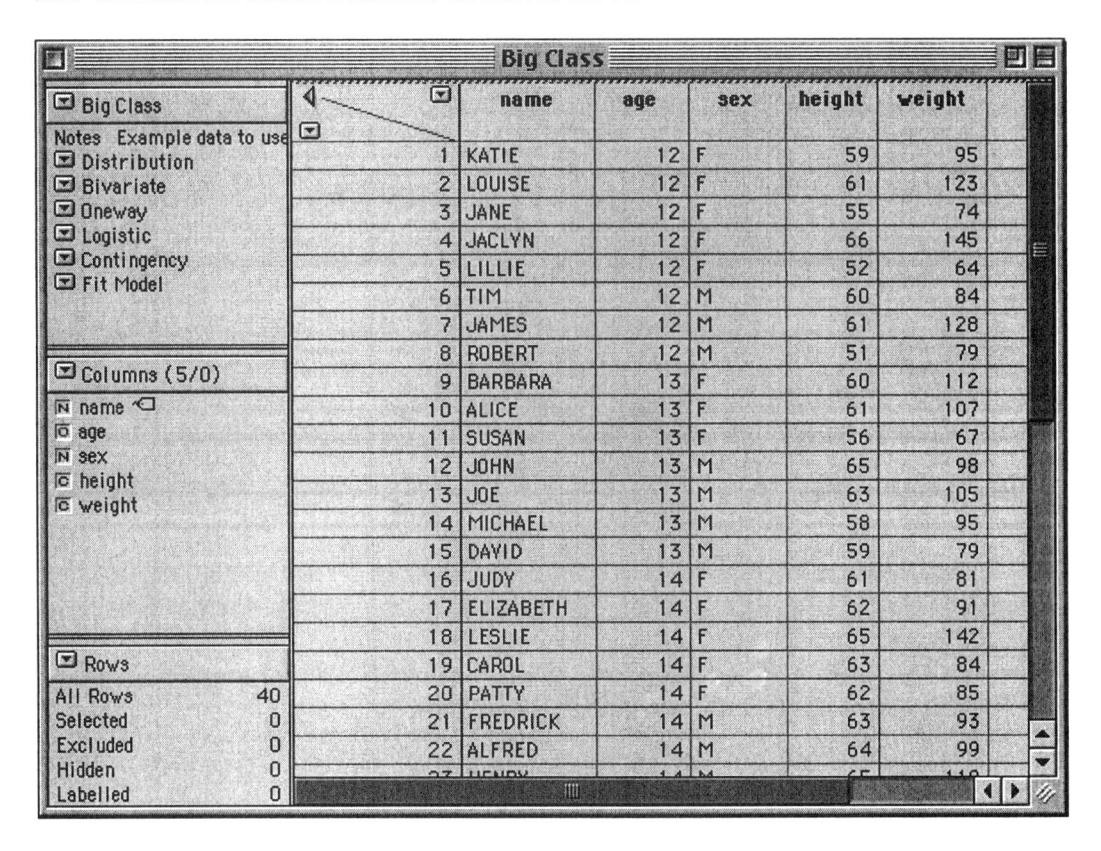

Select a column name (e.g., **age**) in the *Columns Panel*. Notice that the **age** column is highlighted in the *data grid*. Now drag the **age** column name in the *Columns Panel* below **weight**. Notice that the **age** variable has moved to the right of the **weight** variable in the data grid. Press the red triangle on the *Columns Panel* and notice that it duplicates the **Cols** menu on the menu bar. Check out the pop-up menus under the red triangle buttons in the other information panels.

We suggest that the *table information panels* be closed initially until after you become comfortable working with the main menu system and the spreadsheet of data values. The **Preferences** command, discussed next, can be used to close the panels at startup.

0.4.3 Customizing Your Session Environment: Setting Preferences

You may customize your session environment using the **Preferences** command. The **Preferences** command displays a panel with tab pages. The following displays the preferences panel for the Windows environment. Most preferences are also available on the Macintosh.

Example Setting preferences

Let's set preferences to close the Starter Window and the table information panels at startup.

1. To open the preferences window, select **File** ⇨ **Preferences** under Windows, or **Edit** ⇨ **Preferences** on the Macintosh.

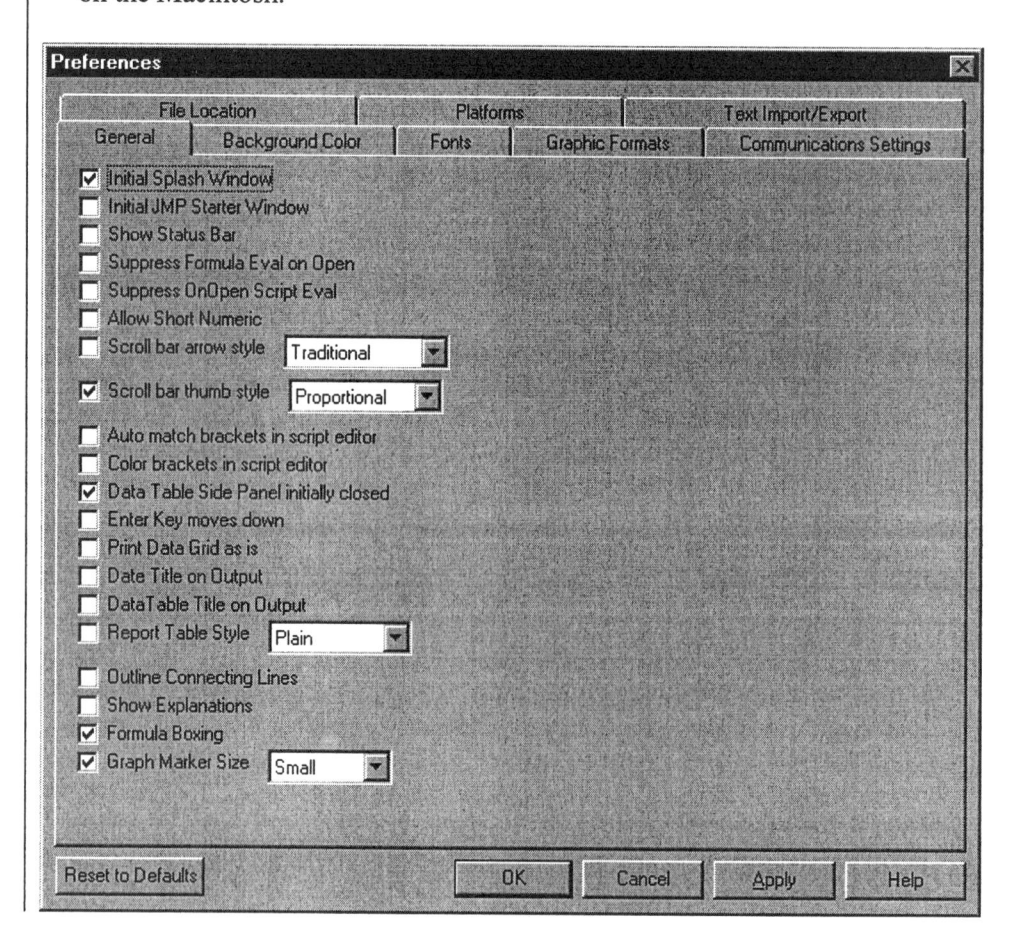

2. Select the **General** tab.

3. Deselect **Initial Starter Window**

4. Select **Data Table Side Panel initially closed** and press **OK**.

Open the preferences panel again and look at the other panel tabs. Most options are either off or on. Check the items that you want or select from a menu of items. Press **Apply** to see the results without closing the **Preferences** window.

0.5 Memorizing and Rerunning Analyses

This section and the next need not be read at this time. They may be more easily understood after reading Chapters 1 and 2. JMP analysis windows are not saved with the data table. However, JMP can memorize a *script* that will reproduce the analysis window, store it with the data table, and replay it later.

0.5.1 Rerunning an Analysis

Let's rerun, or replay, an analysis that has been stored with the **Big Class** data table in the Sample Data folder with the JMP INTRO application.

1. Select **File** ⇨ **Open** ⇨ **Big Class**.

2. Press the disclosure icon in the top left-hand corner of the data grid to open the table information panels. These panels were described more fully in the previous section.

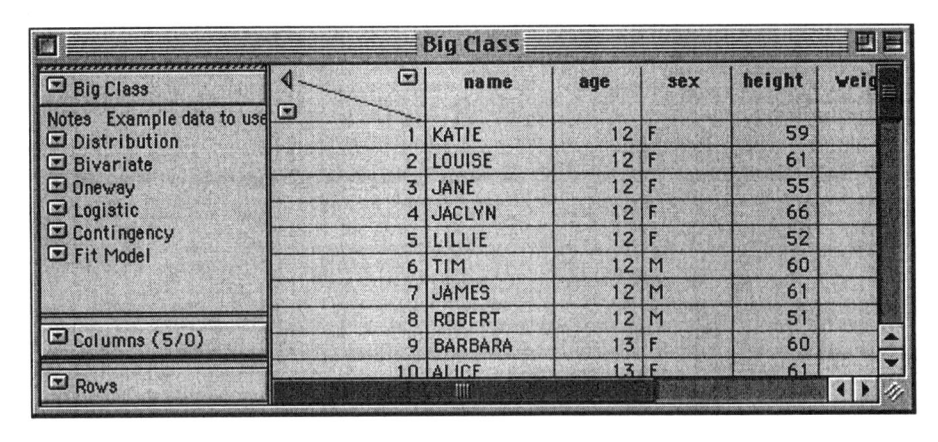

	name	age	sex	height
1	KATIE	12	F	59
2	LOUISE	12	F	61
3	JANE	12	F	55
4	JACLYN	12	F	66
5	LILLIE	12	F	52
6	TIM	12	M	60
7	JAMES	12	M	61
8	ROBERT	12	M	51
9	BARBARA	13	F	60
10	ALICE	13	F	61

3. Click on the red triangle next to **Distribution** in the top panel.

4. Select **Run Script** from the menu that opens.

An analysis window containing the distributions of two variables, **weight** and **age**, is displayed. The *script* that produced the window had been stored with the data table. Other scripts have been stored with the **Big Class** data table as well.

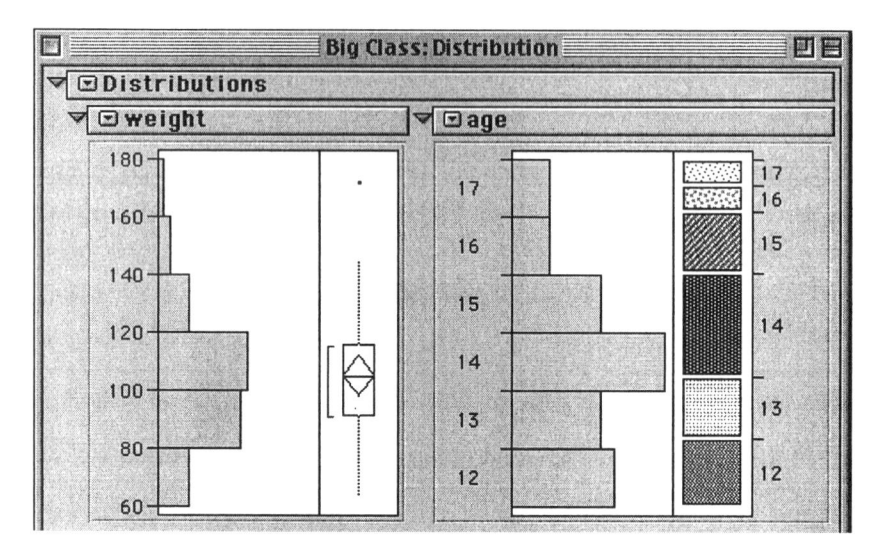

0.5.2 Memorizing an Analysis

How do you tell JMP to memorize and store a *script* that will produce an analysis window? That's almost as easy as replaying a stored *script*. To illustrate, we will first create an analysis window. We do that by replaying a stored *script* and modifying it.

1. Open the JMP data table **Big Class** if it is closed.

2. Press the red triangle next to **Oneway** on the Table panel that is to the left of the data grid, and select **Run Script** from the menu that opens.

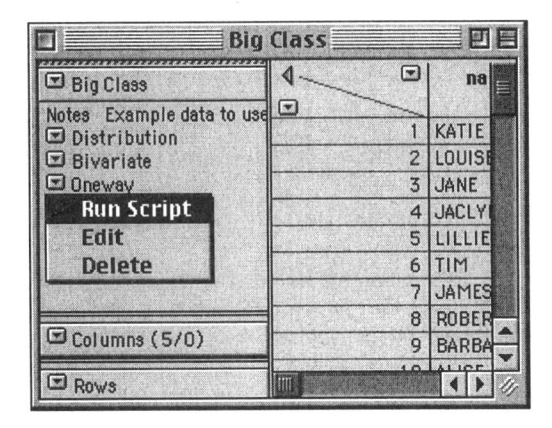

Let's modify this plot to get side-by-side boxplots of the heights—one for the girls and one for the boys.

3. Press the red triangle next to **Oneway Analysis of height By sex** and deselect **Means/Anova/t-Test** on the menu that opens.

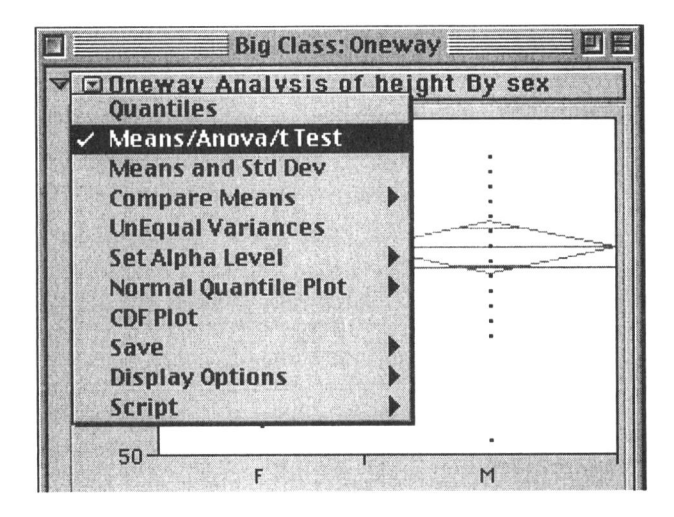

4. Press the red triangle again and select **Display Options** ⇨ **Box Plots**.

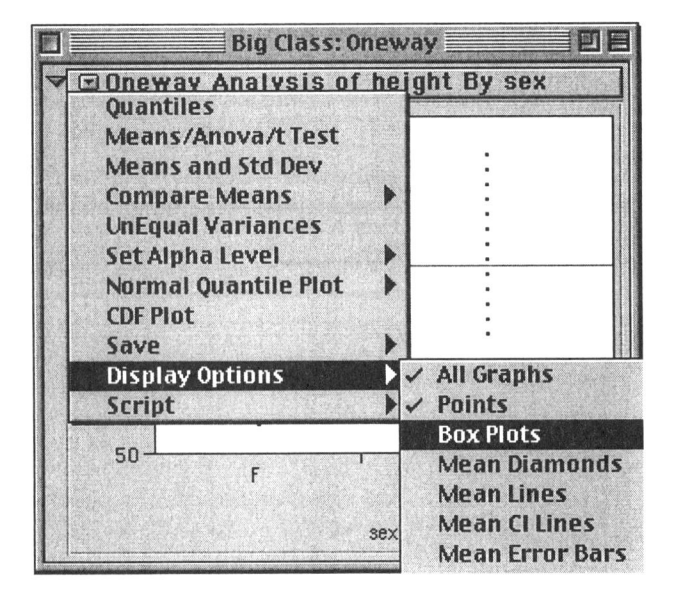

5. Press the red triangle again and deselect **Display Options** ⇨ **Points**.

To have JMP memorize a script that will produce this analysis window, simply

6. Press the red triangle again and select **Script** ⇨ **Save Script to Datatable**.

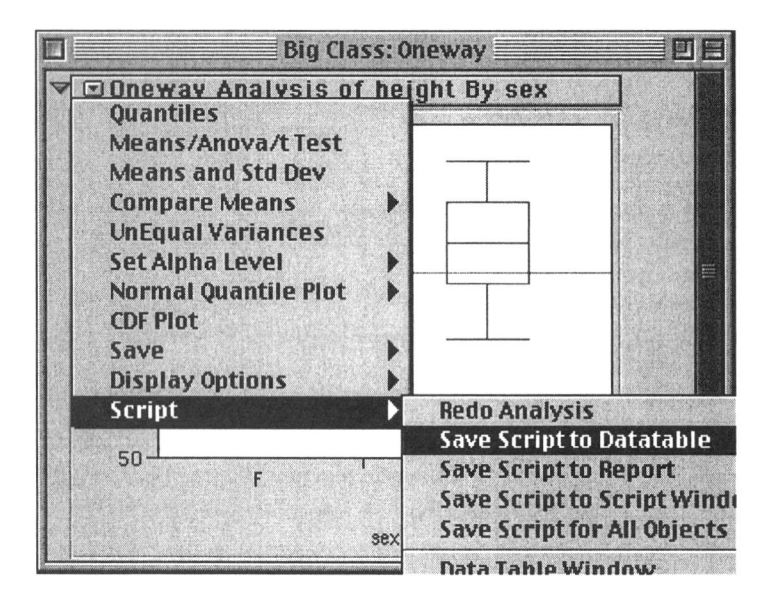

That's all there is to it. Go to the **Big Class** data table and look at the Table panel. Notice the new script called **Oneway 2**. That's your script.

7. To rename the script, double-click on the name of the script **Oneway 2**.

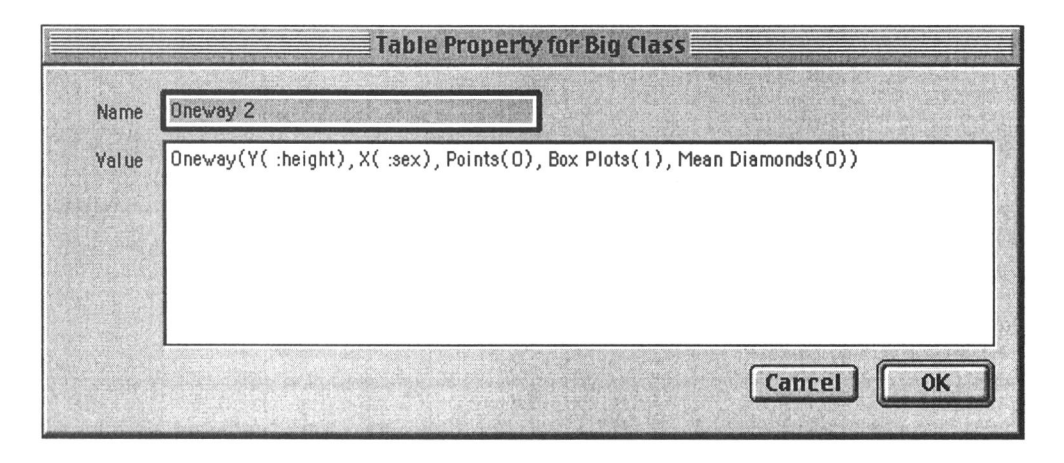

8. Type `My First Script` in the **Name** box and click **OK**.

Remarks

Scripts are written in a tremendously powerful language called the *JMP Scripting Language (JSL)*. JSL is a complete programming language. In addition to facilitating the recreation and storage of analyses, JSL can be used to more easily perform routine tasks and to extend JMP beyond its already broad capabilities. An example of the language itself can be seen in the **Value** box in the figure above.

A collection of sample scripts is included in the Sample Scripts folder in the subdirectory containing the JMP INTRO application. To run a script, open it and select **Run Script** from the **Edit** menu. Let's run one of the included scripts, **Confidence.jsl**, that we will use in Chapter 6 to illustrate the meaning of a confidence interval.

1. Select **File** ⇨ **Open** from the menu bar.

2. Select **Confidence.jsl** from the Sample Scripts folder.

3. Select **Edit** ⇨ **Run Script**. An interactive window opens.

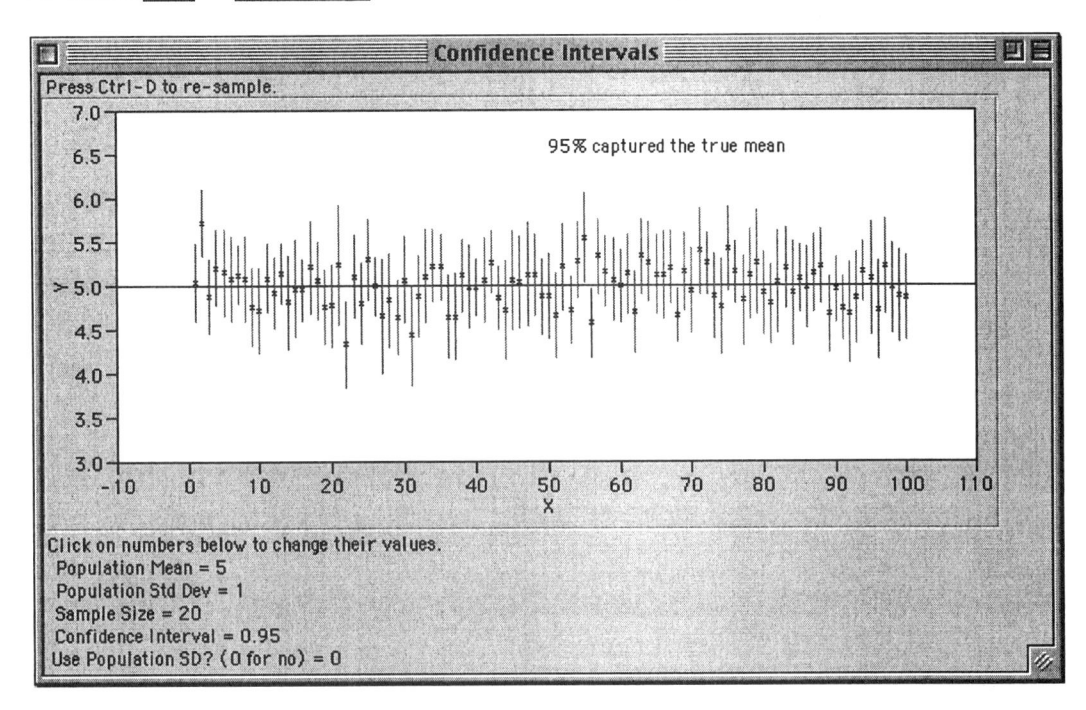

0.6 Enhancing Reports and Plots

For most projects, printed copies of JMP report windows, which contain the statistical analyses and graphics, are all that is needed. Occasionally, you may wish to modify these reports by reformatting calculations, resizing plots, changing scales, removing parts of the reports, or adding annotation. Sometimes you may wish to copy parts of the reports to your favorite word processor. You can do this in JMP INTRO.

Here are some ways that you can control the appearance and content of your reports, plots, and charts.

Report Windows
- Parts of a report can be closed and not printed.
- Titles and subtitles of a report can be changed.
- Report tables and calculations can be reformatted.
- Editable notes can be added to a text report or plot using the *Annotate* tool.

- Parts of a report can be copied to a word processor using the *Selection* tool.
- Text styles of the reports can be specified using the Font tab on the **Preferences** panel discussed in Section 0.4.3.

Plots and Charts
- Individuals in plots and the data table are linked. Select data points or frequency bars and notice that the corresponding individuals in the data table or other plots are highlighted as well.
- Plots can be resized by selecting an edge or the lower-right-hand corner and dragging.
- Axes can be scrolled by moving the hand tool over the numbers.
- Axes' names can be changed by double-clicking on the name.
- Many aspects of an axis can be customized by double-clicking on the values. Tick marks, range, and the increment between numbers can be modified; reference lines can be added.
- Formats for graphics (jpeg, png, metafile) can be specified using the Graphics Format tab on the **Preferences** panel discussed in Section 0.4.3.

Some of these features are presented in Chapters 1 and 2; others are presented here. You should read and perform the analyses in Chapters 1 and 2 before proceeding with the rest of this section.

0.6.1 Report Windows

As you have noticed by now, JMP reports are organized in hierarchical outline. Each level of the outline has a blue *disclosure/closure* icon. You click these buttons to open and close the corresponding analysis or report at that outline level. When an outline level is closed, only the open levels are printed. We will briefly show some of the other ways that JMP allows you to control and enhance the content of your report windows.

First, we create a report window using one of the scripts stored with the JMP data table **Big Class**.

1. Open the data table **Big Class**.

2. Press the red triangle next to **Oneway** on the Tables panel to the left of the data grid.

3. Select **Edit** ⇨ **Run Script** to create the report window **Big Class: Oneway**.

You can change the title of the report and close reports in the window.

4. Double-click on the report title **Oneway Analysis of height By sex** and type `Are Height and Gender Related?`.

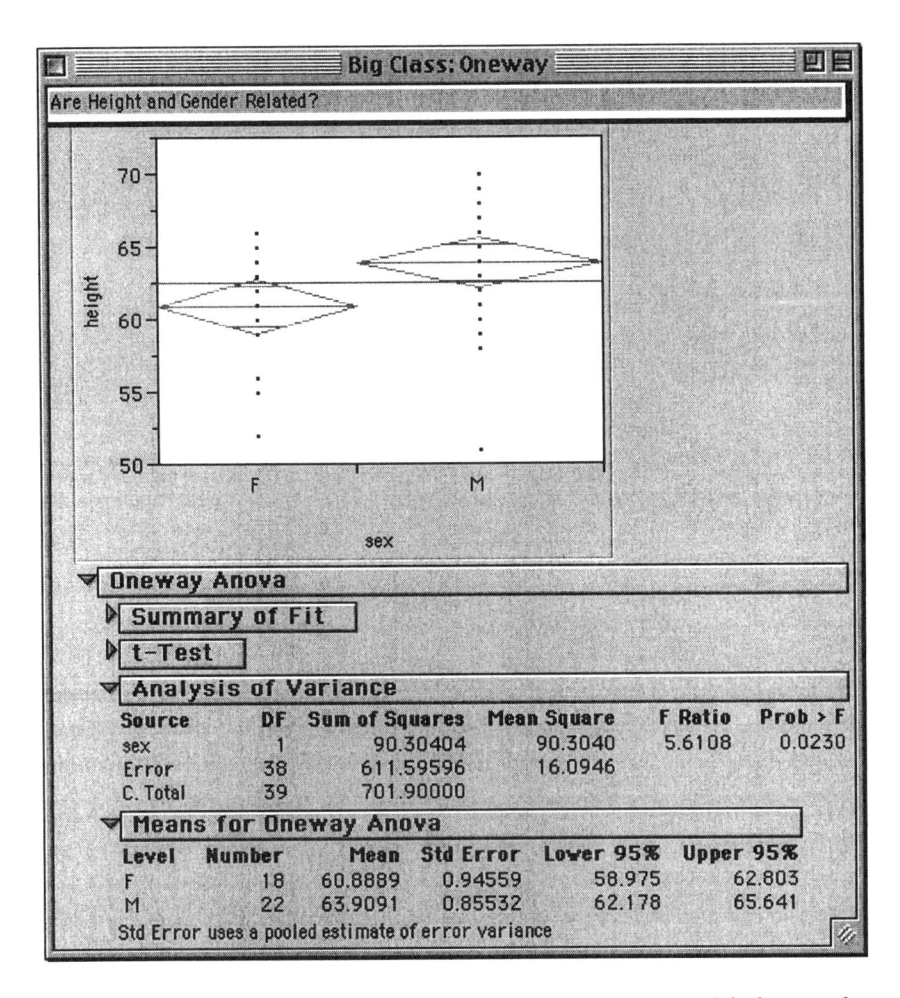

![Big Class: Oneway window showing scatter plot of height by sex with means diamonds, and Oneway Anova report with Analysis of Variance table and Means for Oneway Anova table]

Analysis of Variance

Source	DF	Sum of Squares	Mean Square	F Ratio	Prob > F
sex	1	90.30404	90.3040	5.6108	0.0230
Error	38	611.59596	16.0946		
C. Total	39	701.90000			

Means for Oneway Anova

Level	Number	Mean	Std Error	Lower 95%	Upper 95%
F	18	60.8889	0.94559	58.975	62.803
M	22	63.9091	0.85532	62.178	65.641

Std Error uses a pooled estimate of error variance

5. Press the *disclosure* icon next to the **Summary of Fit** title bar to close that report.

6. Press the *disclosure* icon next to the **t-Test** title bar to close that report also.

You can remove the columns containing the 95% confidence intervals for the means in the **Means for Oneway ANOVA** report and change the format of the sum of squares values in the **Analysis of Variance** report.

7. Double-click on one of the values in the **Sum of Squares** column of the **Analysis of Variance** report. A format panel appears.

8. Enter **2** in the **Decimal box** and press **OK**.

9. Context-click in the **Means for Oneway Anova** report (right-click under Windows and crtl-click on a Macintosh).

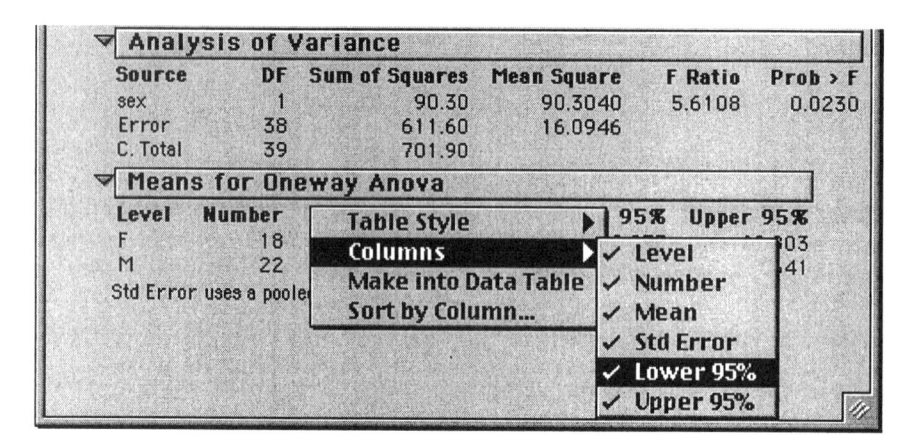

10. Deselect **Columns** ⇨ **Lower 95%**.

11. Context-click on that report again and deselect **Columns** ⇨ **Upper 95%**.

You can annotate the report. Add a legend for **sex** next to the plot.

12. Select the Annotate tool from the **Tools** menu.

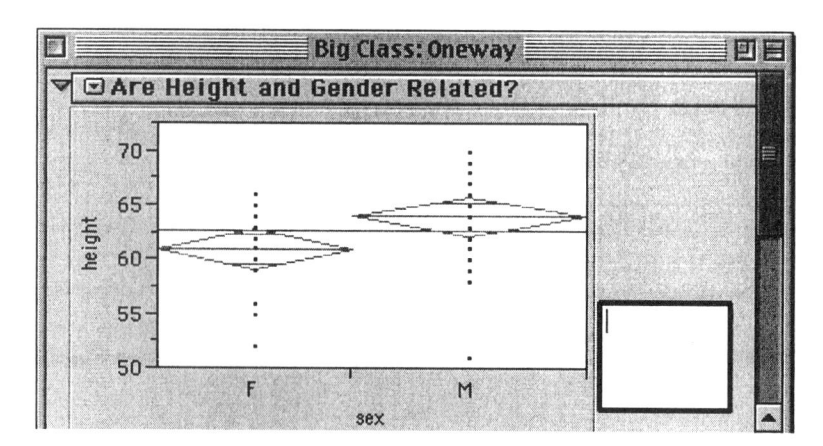

13. Draw a rectangle next to the plot and type `F=Female` and `M=Male`.

Compare the resulting report window with the first figure in this section and note that neither the reports for **Summary of Fit** and **t Test** nor their titles will be printed.

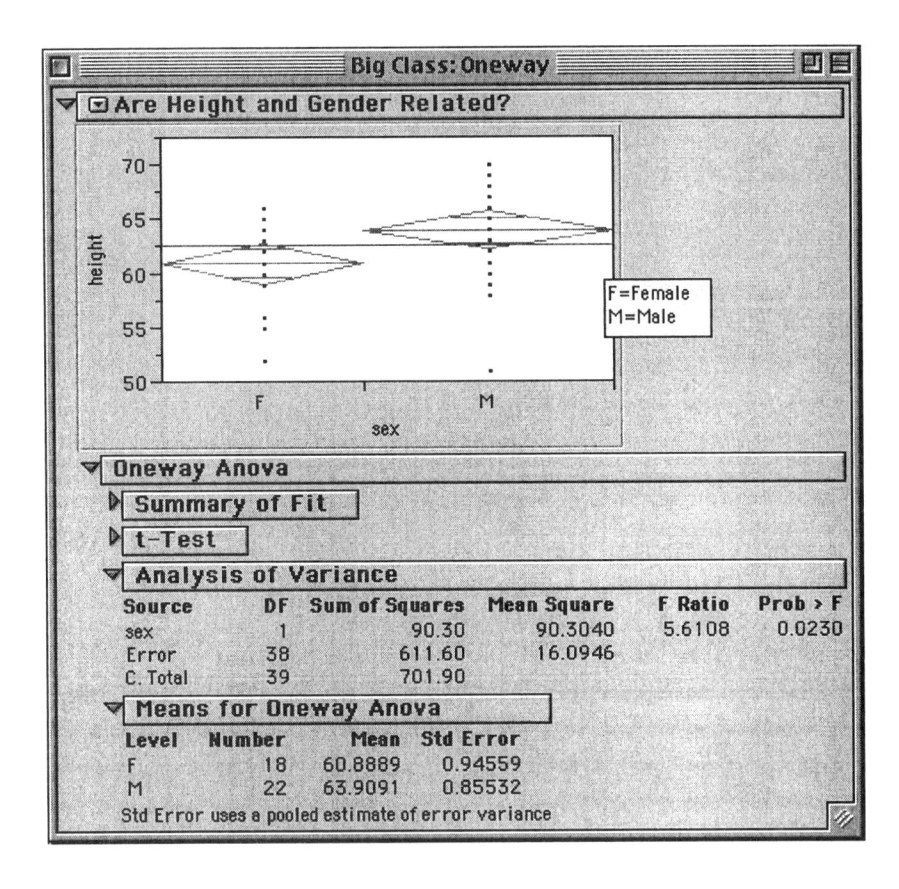

Suppose that instead of reformatting the report window, you wish to copy a portion of it into your favorite word processor.

1. Select the **Area Selection Tool** (a dotted rectangle on a Macintosh and a fat plus sign under Windows).

2. Select the **Analysis of Variance** report.

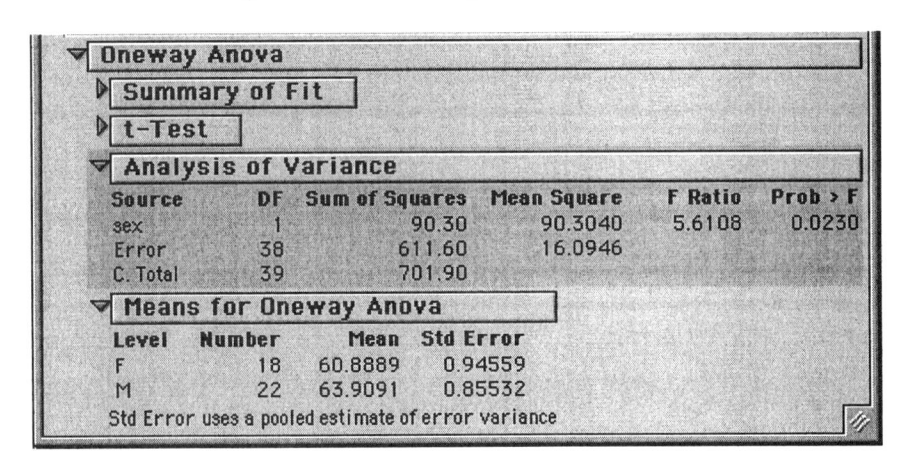

3. Select **Edit** ⇨ **Copy** to copy the report to the clipboard. Then, simply paste it into your word processor.

0.6.2 Plots and Charts

Axes can be customized in JMP by *double-clicking* or *context-clicking* on either the axis name or the axis values. *Double-click* on an axis name to change it. Double-click on axis values and the Axis Specification dialog appears.

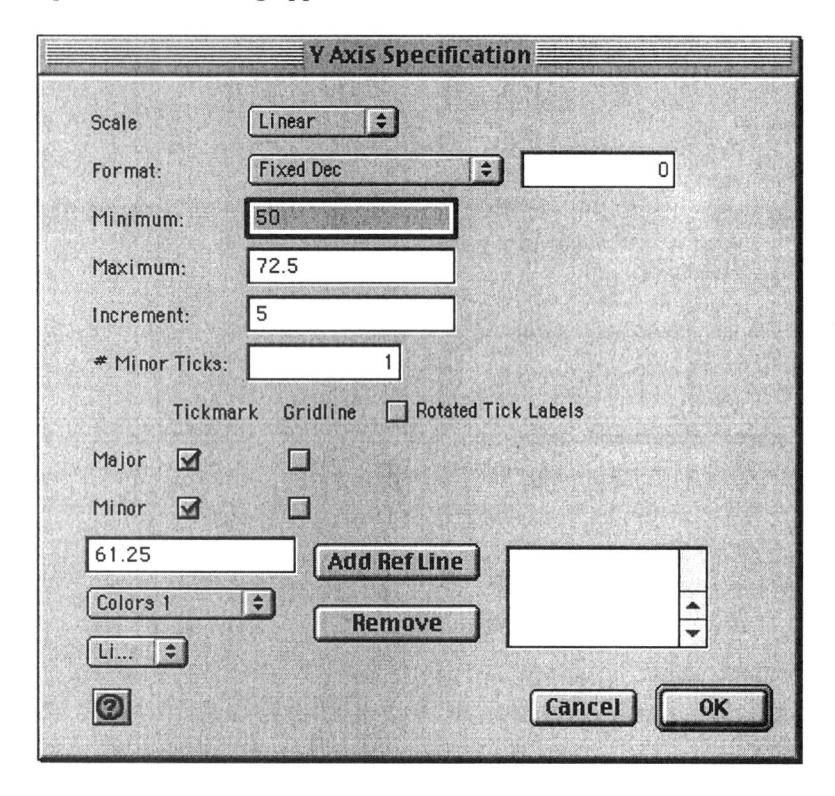

Context-click (right-click under Windows and ctrl-click on a Macintosh) to see additional options such as font management and text rotation.

Remark

The professional version of JMP offers considerably more powerful tools for capturing and customizing the results of a JMP analysis. Creating a final report is both quick and easy with the **Journal** and **Layout** commands.

0.7 Exercises

1. Survey of Study Habits and Attitudes (SSHA) scores. The file **ex02_016.txt** on the IPS CD-ROM contains SSHA scores for first-year college students. Put the data in a JMP data table and save it for later use.

2. Ethnicity of U.S. college undergraduates. Exercise 1.11 of IPS has data on the ethnicity of undergraduate students in U.S. colleges in 1997. Create a JMP data table to hold this data. Use the

data table **fg01_001.jmp** stored in the Sample Data folder and discussed in Section 1.1 in Chapter 1 as a guide.

3. Is wine good for your heart? Table 2.2 of IPS gives data on yearly wine consumption and yearly deaths from heart disease. The data are stored in the text file **ta02_002.txt** on the IPS CD-ROM. Use this file to create a JMP data table and save the data table for later use.

4. Does calcium intake affect blood pressure? You imported data from an experiment on blood pressure in Section 0.2.1, Example 7.19. The resulting data table contains readings before and after a 12-week period and a variable **Decrease** that is supposed to represent the decrease in blood pressure reading. Check this by creating a new variable **Diff** which equals the before (Begin) reading minus the after (After) reading. Compare it to the variable **Decrease**.

Chapter 1

Looking at Data—Distributions

This chapter examines the distribution of a variable. Appropriate graphical and semigraphical methods for displaying distribution are presented; numbers that described the location and spread of a distribution are discussed; and models (density curves) for distributions are described along with methods of assessing the quality of their fit.

1.1 Displaying Distributions with Graphs

All graphs and statistical computations for this chapter, with one exception, are performed in the **Distribution** platform of the **Analyze** menu. There is no need to tell JMP that you want a bar chart or a histogram. JMP automatically produces the appropriate graph and numeric summaries depending on the type of variable. For a categorical or nominal variable, JMP produces bar charts and a frequency table. For a continuous, quantitative variable, it produces a histogram and calculates the five-number summary, the mean and standard deviation.

1.1.1 Categorical Variables: Bar Graphs

IPS Figure 1.1(a)   Educational level of 30-something young adults

Figure 1.1(a) of IPS displays the educational level of 30-something young U.S. adults. The JMP data table **fg01_001.jmp** in the Sample Data folder contains the educational level data.

1. To open the data table, select **File** ⇨ **Open** in the menu bar.

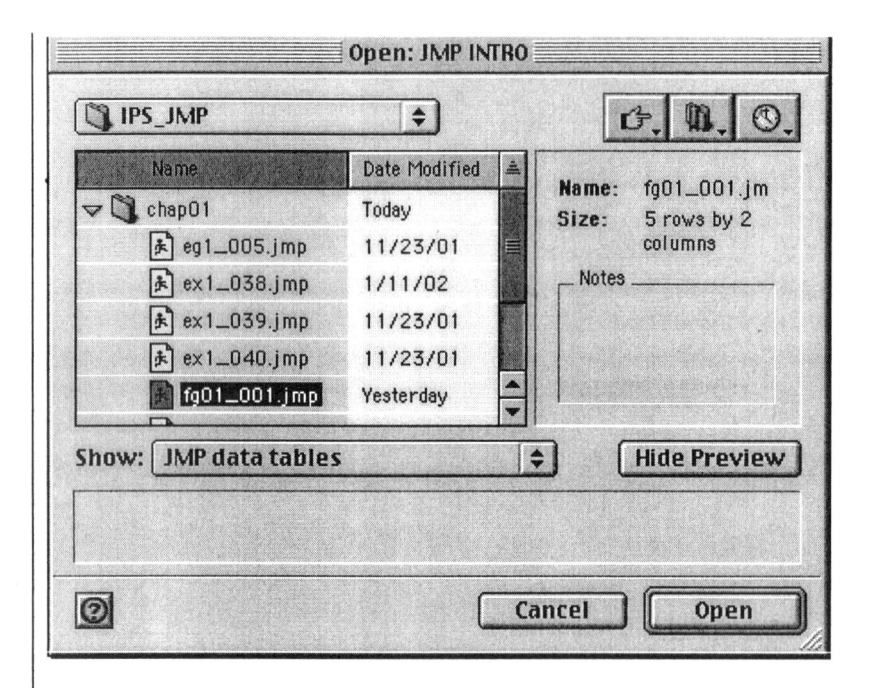

2. Select the data table **fg01_001.jmp** and press **Open**.

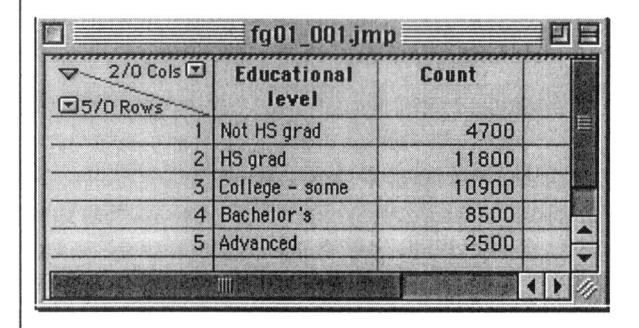

To display the distribution of educational levels, use the *Distribution* analysis platform.

3. Select **Analyze** ⇨ **Distribution** from the menu bar.

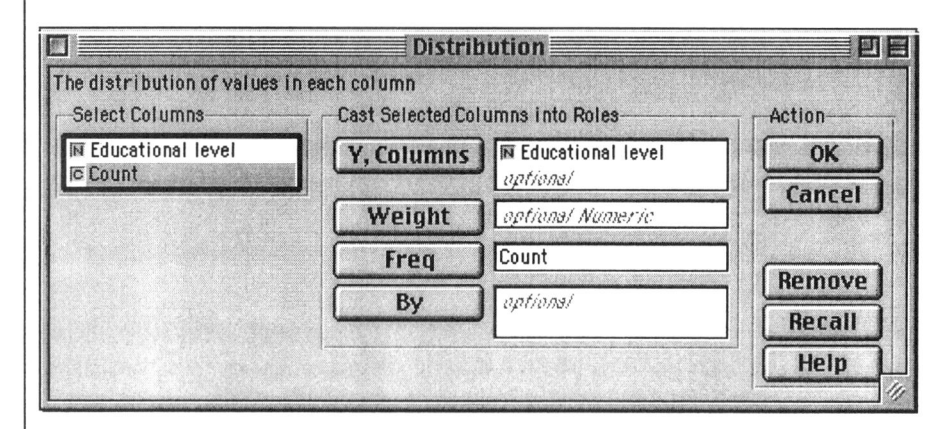

4. Select the column **Educational level** and press the **Y, Columns** button.

5. Select the column **Count** and press the **Freq** button.

6. Select **OK**.

JMP displays a frequency table, a regular bar chart, and a stacked bar chart to quickly compare the sizes of the five education categories.

```
fg01_001.jmp: Distribution
Distributions
  Educational level
```

Level	Count	Prob
Advanced	2500	0.06510
Bachelor's	8500	0.22135
College - some	10900	0.28385
HS grad	11800	0.30729
Not HS grad	4700	0.12240
Total	38400	1.00000
5 Levels		

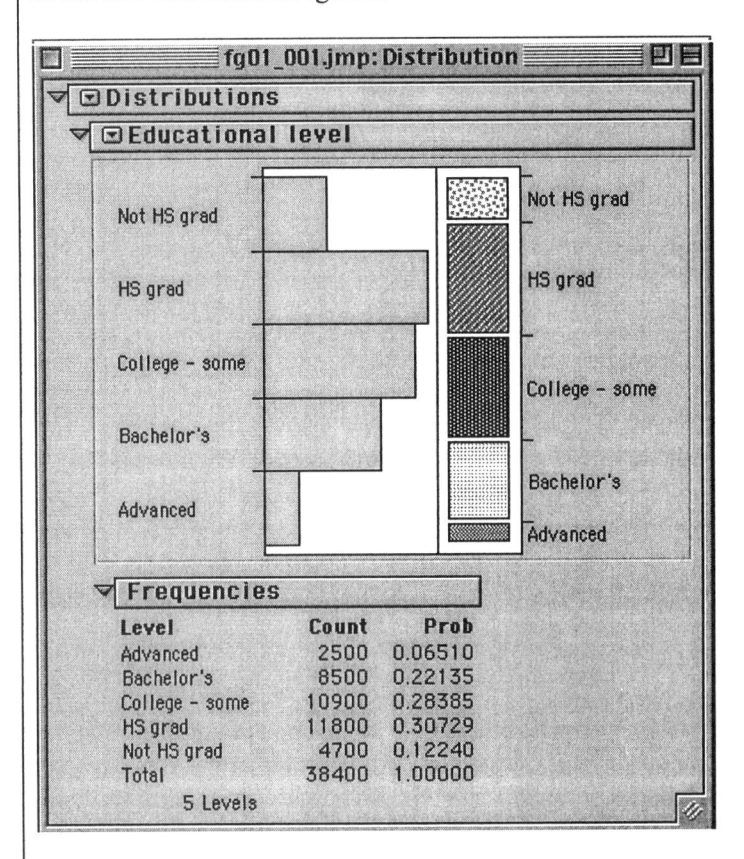

To display the regular bar chart horizontally as in Figure 1.1(a) of the text,

7. Click on the red triangle in the title of the **Educational level** report.

8. Select **Display Options** ⇒ **Horizontal Layout** from the menu that opens.

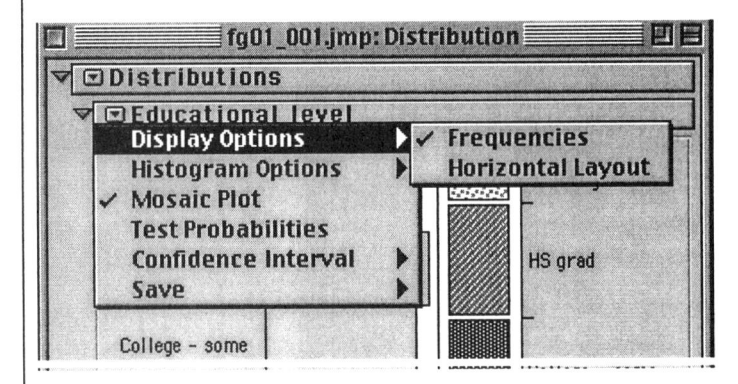

Notice that the names of the education categories are easier to read in vertical layout. This is why it is the default layout.

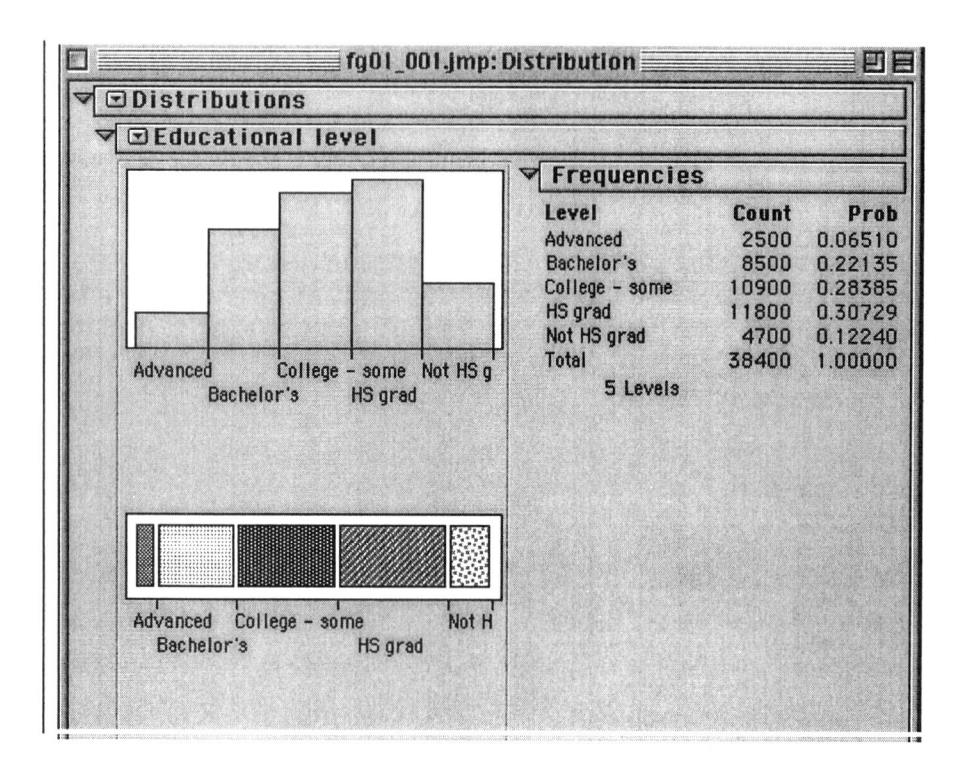

![JMP Distribution window for Educational level showing histogram and frequency table]

```
fg01_001.jmp: Distribution
Distributions
  Educational level

Frequencies
Level         Count    Prob
Advanced       2500   0.06510
Bachelor's     8500   0.22135
College - some 10900  0.28385
HS grad       11800   0.30729
Not HS grad    4700   0.12240
Total         38400   1.00000
5 Levels
```

1.1.2 Quantitative Variables: Histograms and Stemplots

For quantitative variables, the values of the variable must be grouped together to see the distribution of the values. A *histogram* and a *stemplot* are used to display the distribution of values for a quantitative variable. Both are extremely easy to produce in JMP.

IPS Example 1.5 Supermarket spending

Example 1.5 of IPS gives the amounts spent by 50 consecutive shoppers in a supermarket. Let's create a histogram to examine the distribution of the variable **Amount spent**.

The JMP data table **eg01_005.jmp** in the Sample Data folder contains the 50 amounts spent.

1. Select **File** ⇨ **Open** in the menu bar to open the data table.

Looking at Data—Distributions 43

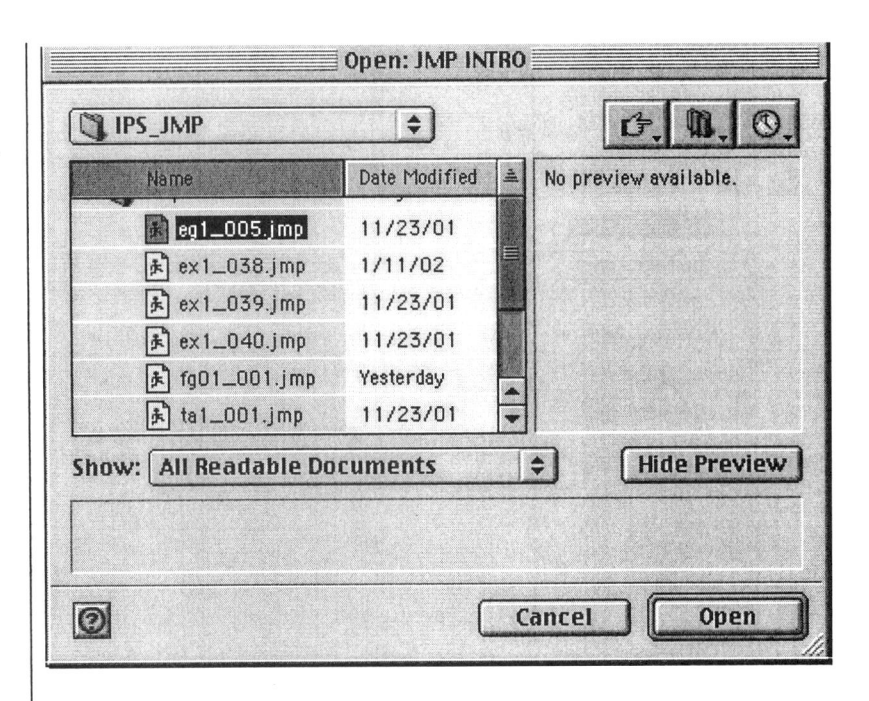

2. Select the file **eg01_005.jmp** and press **Open**.

To display the distribution of amounts spent by these 50 shoppers, use the first analysis platform **Distribution**.

3. Select **Analyze** ⇨ **Distribution**.

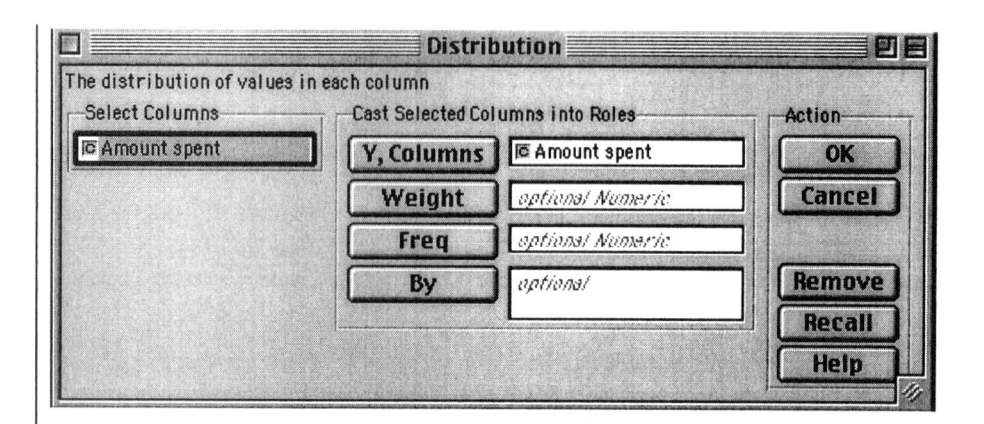

4. Select the column **Amount spent** and press the buttons **Y, Columns** and **OK**.

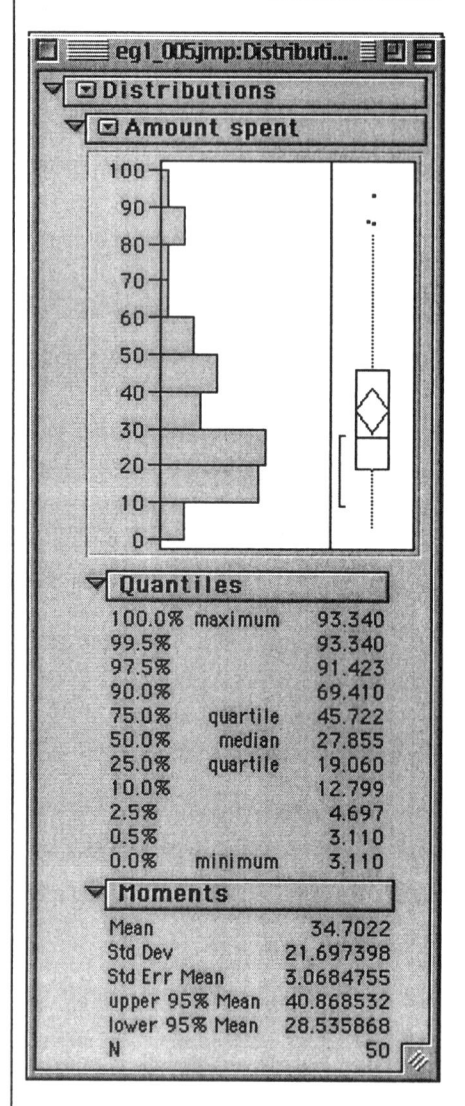

Notice that the amounts spent are skewed toward higher values with an average of about $25 to $30. By default, JMP displays the histogram values on a vertical scale. You may wish to see the more familiar horizontal layout.

Looking at Data—Distributions

5. Click on the red triangle in the title of the **Amount spent** report.

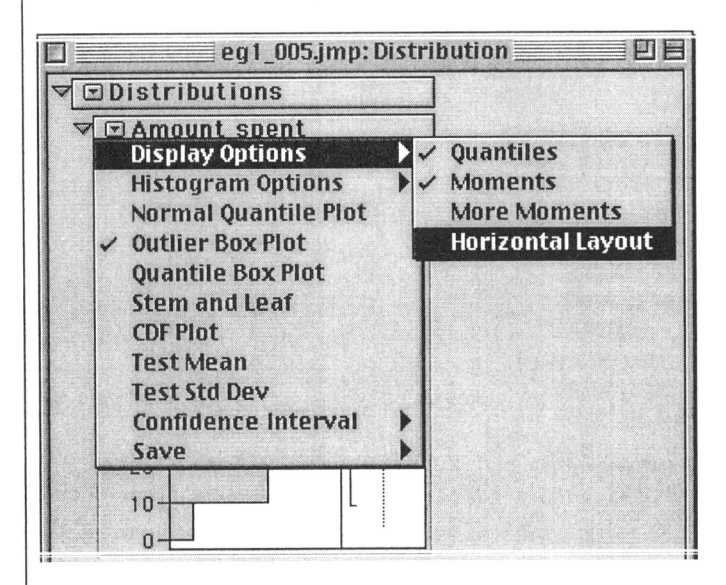

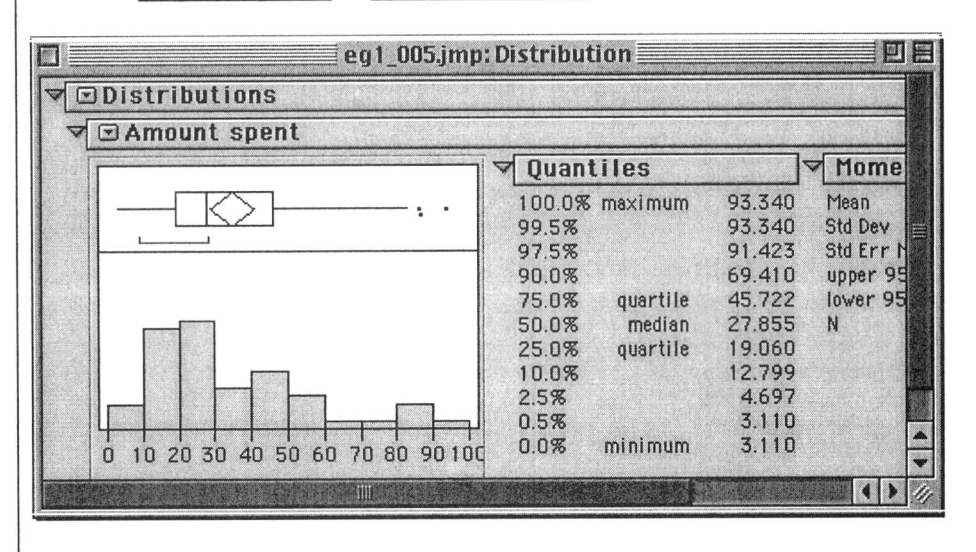

6. Select **Display Options** ⇨ **Horizontal Layout**.

Adjusting the Number of Bars. Usually, we investigate whether another choice of interval width and starting point provides more insight into the overall pattern or into the deviations from the pattern. That is very easy to do in JMP using the *hand tool*. First, switch back to the default vertical layout.

1. a. Click on the red triangle in the title of the **Amount spent** report.
   b. Deselect **Display Options** ⇨ **Horizontal Layout** to uncheck the horizontal layout option.

2. Now select the hand tool from the **Tools** menu.

3. Move the mouse in the direction of the frequency scale (side-to-side for the default vertical layout). Notice that the number and width of the bars, or intervals, changes.

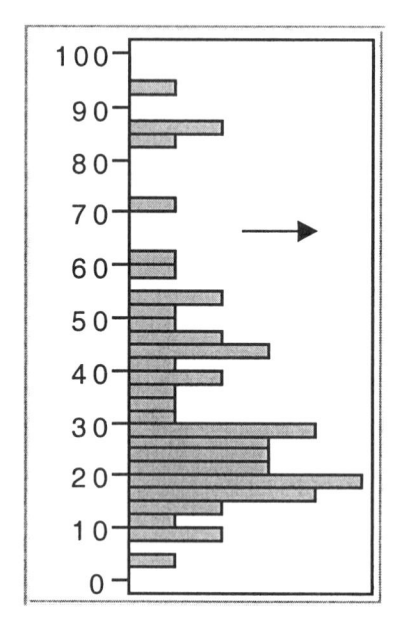

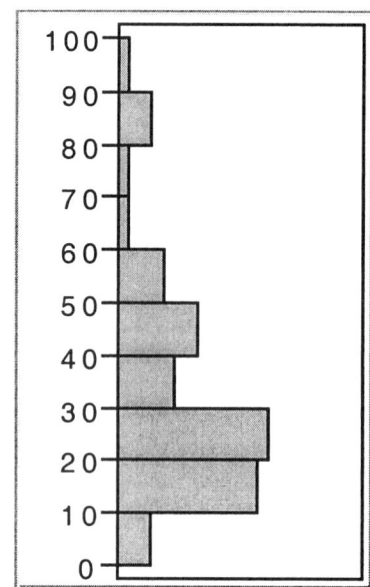

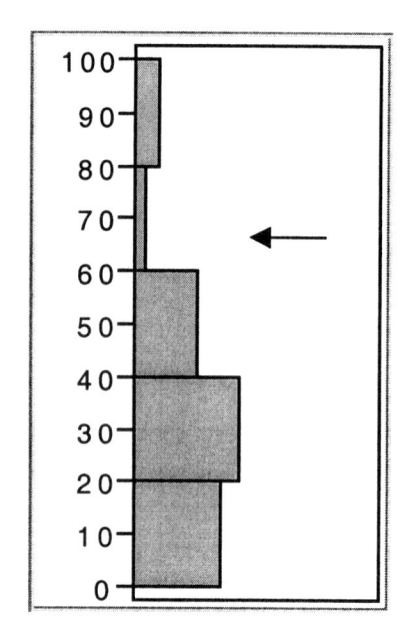

4. Move the mouse up and down along the values axis. Notice that the starting point of the intervals changes.

5. Experiment with this. Compare the histograms drawn with intervals of width 10 and width 20.

Stemplots

To display a stemplot,

1. Click on the red triangle in the title of the report **Amount spent.**

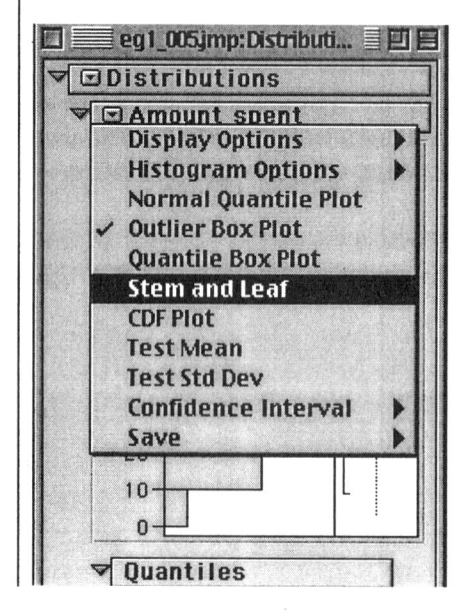

2. Select **Stem and Leaf** from the pop-up menu.

You may need to scroll down to the bottom of the report to see the stemplot. Or you can click on the blue icons (closure/disclosure buttons) for **Quantiles** and **Moments** to close those reports.

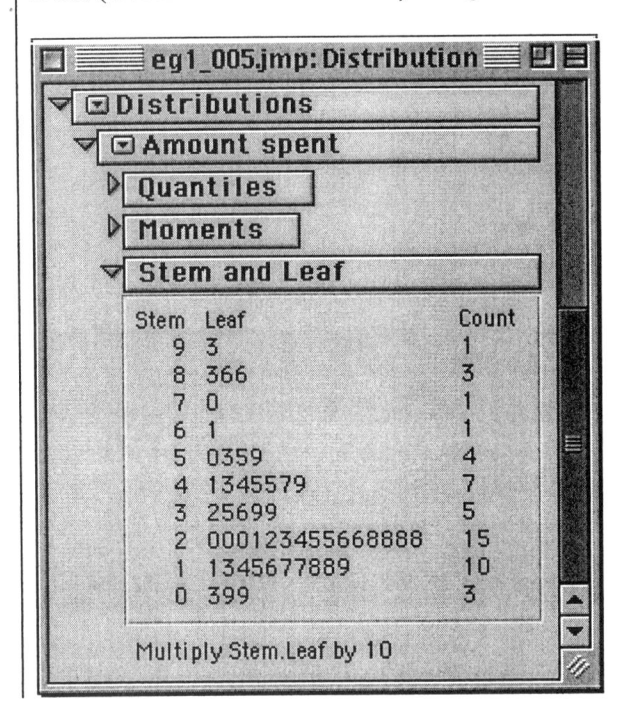

```
eg1_005.jmp: Distribution
▽ Distributions
  ▽ Amount spent
    ▷ Quantiles
    ▷ Moments
    ▽ Stem and Leaf
      Stem  Leaf              Count
        9   3                    1
        8   366                  3
        7   0                    1
        6   1                    1
        5   0359                 4
        4   1345579              7
        3   25699                5
        2   000123455668888     15
        1   1345677889          10
        0   399                  3
      Multiply Stem.Leaf by 10
```

1.1.3 Time Plots

To create a time plot in JMP IN or JMP, use the **Time Series** command in the **Analyze** menu. To create a time plot in JMP INTRO, use the *Chart* platform on the **Graph** menu. We illustrate the steps for JMP INTRO.

IPS Table 1.1 Passage time of light

Table 1.1 of IPS gives data on Simon Newcomb's 66 measurements of the passage time of light. Let's plot the data in the order in which the measurements were taken. The text file **ta01_001.txt** on the IPS CD-ROM contains the measurements in the order in which they were recorded.

1. Import the file **ta01_001.txt** into a JMP data table (see Section 0.2.1 in Chapter 0 for details on importing data) and name the column of data values **Passage Time**.

48   Looking at Data—Distributions

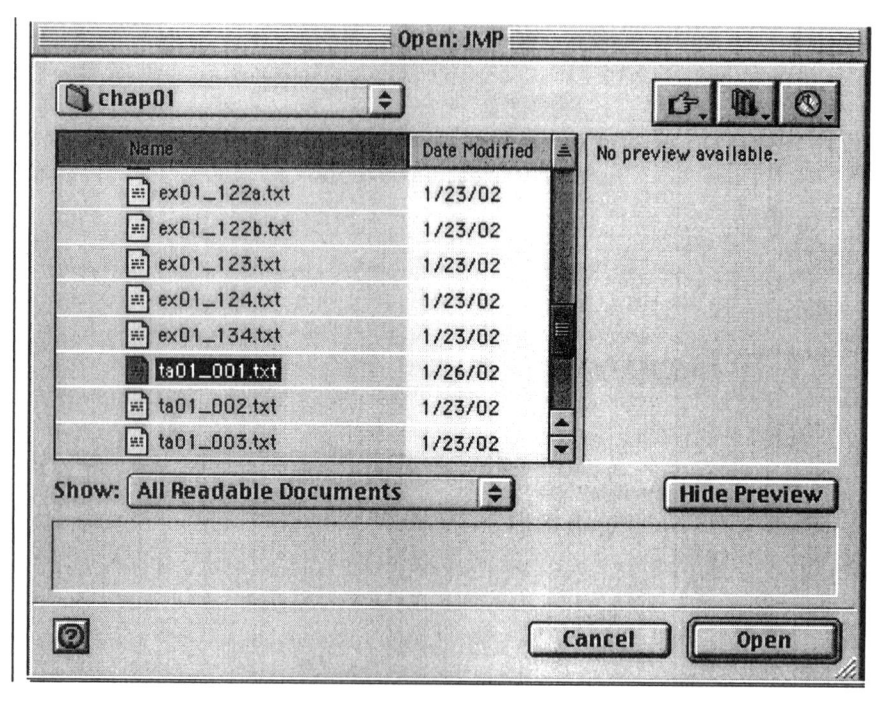

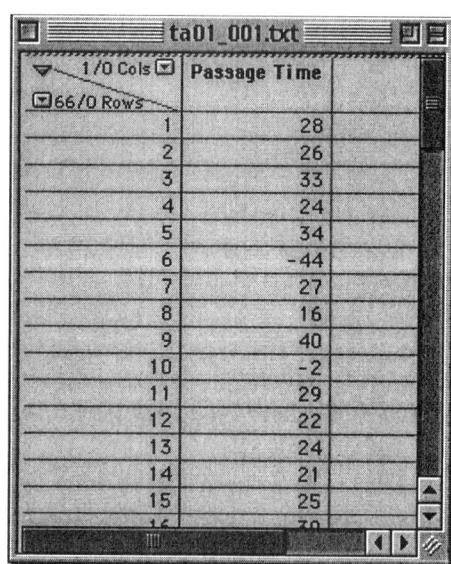

Save the resulting JMP data table for later use. Name it **ta01_001.jmp**.

2. **File** ⇨ **Save**.

3. Enter `ta01_001.jmp` in the **Name** field of the **Save** dialog and press **Save**.

4. Select **Graph** ⇨ **Chart**.

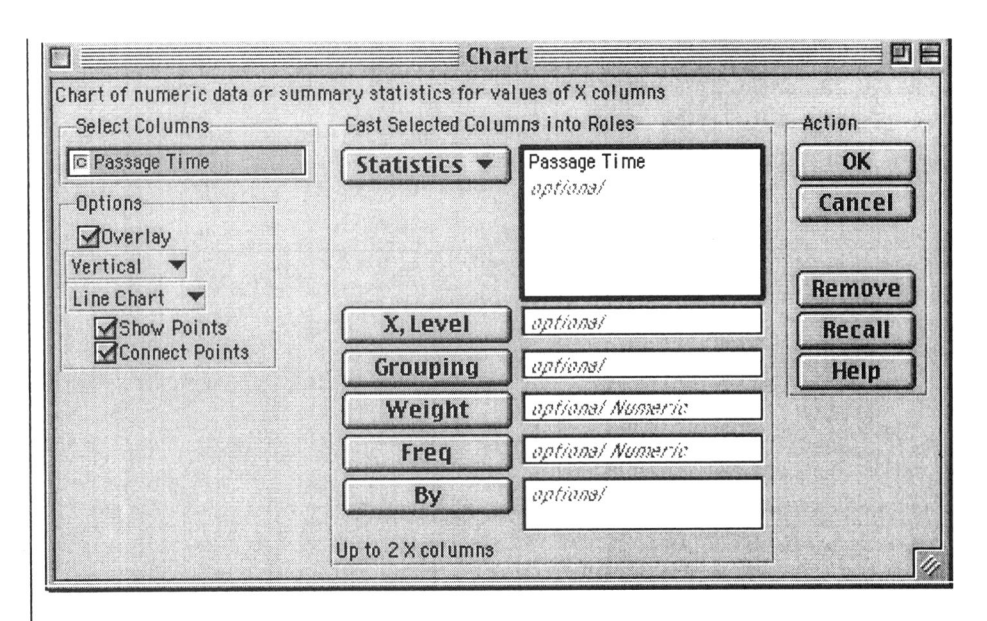

5. Select **Passage Time** from the list of columns. (It's the only variable there.)

6. Select **Statistics** ⇨ **Data**.

7. Select **Line Chart** from the pull-down menu titled **Bar Chart** under **Options**.

8. Click **OK**.

9. Click on the red triangle in the **Chart** title bar.

10. Deselect **Show Level Legend** to remove the legend for the X-axis.

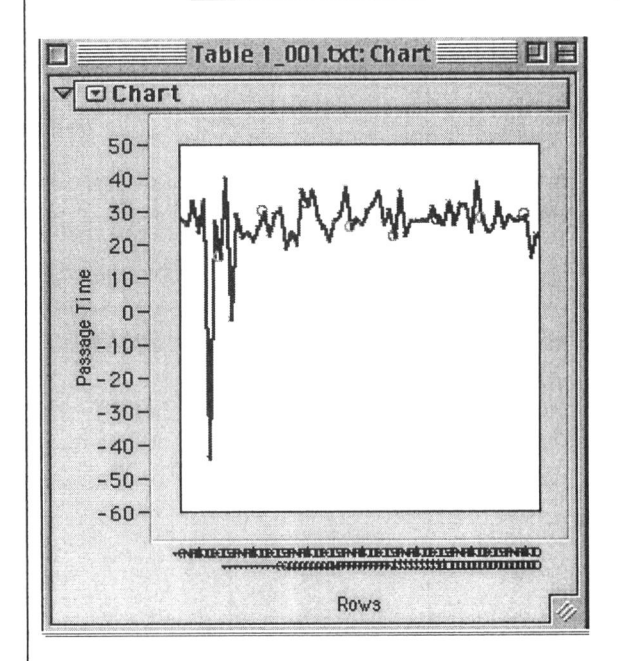

Resizing the Plot and Changing Marks on the X-axis. You may wish to enlarge the graph, change its aspect ratio, or modify the axis display. These are easy to do in JMP.

1. Place the mouse near the lower-right-hand corner of the graph. The mouse will change to a two-headed offset arrow. Move the arrow to resize the graph.

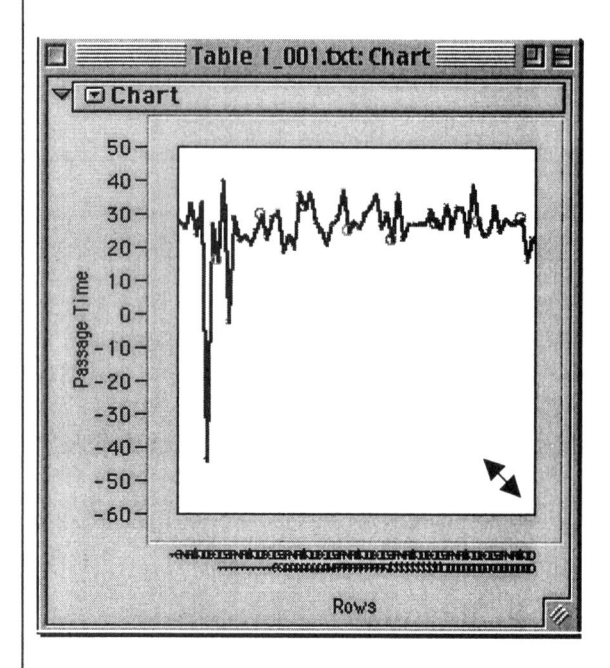

2. Double-click anywhere in the list of values on the X-axis.

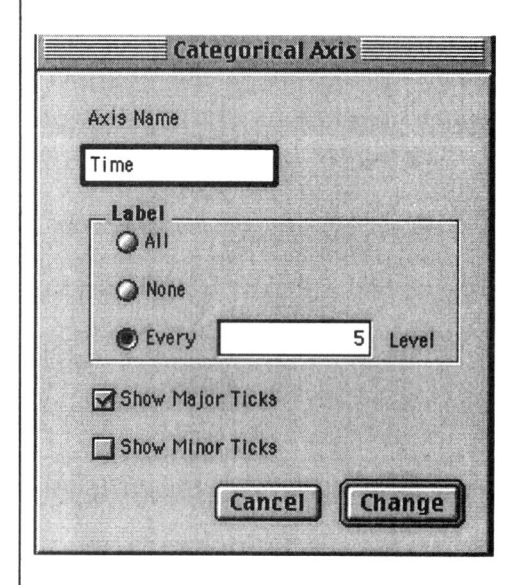

3. Enter **Time** as the **Axis Name**.

4. Enter **5** to label every 5th level.

5. Select **Show Major Ticks**.

6. Press **Change**.

Looking at Data—Distributions 51

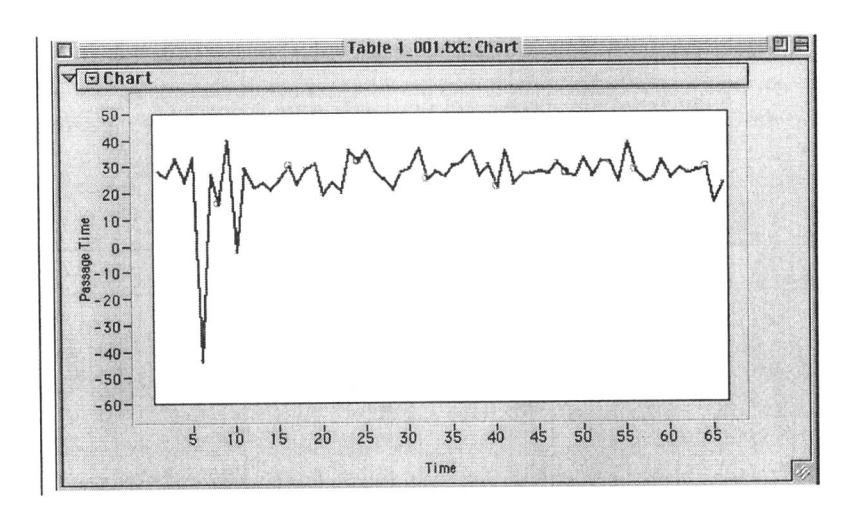

Time as a Variable. Sometimes the calendar times of the data are explicitly recorded. This is the case in Exercise 1.38. The text file **ex01_038.txt** contains a variable for time **Year** and a response variable **Death Rate**. We want to plot **Death Rate** versus **Year**.

IPS Exercise 1.38 U.S. motor vehicle deaths

Import the file **ex01_038.txt** on the IPS CD-ROM into a JMP data table (see Section 0.2.1 in Chapter 0 for more details) and name the columns **Year** and **Death Rate,** respectively.

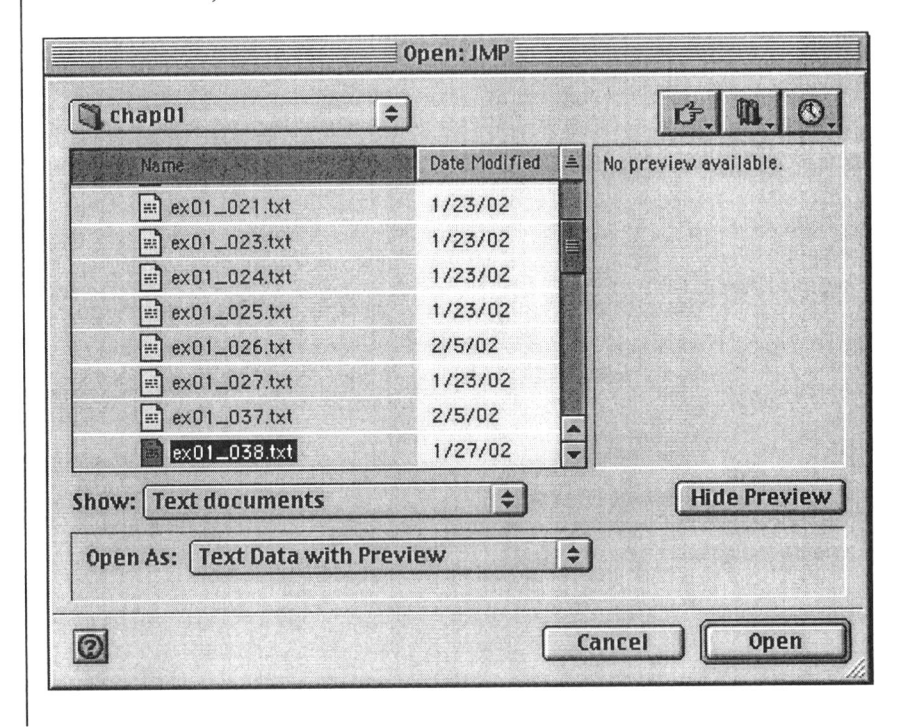

52   Looking at Data—Distributions

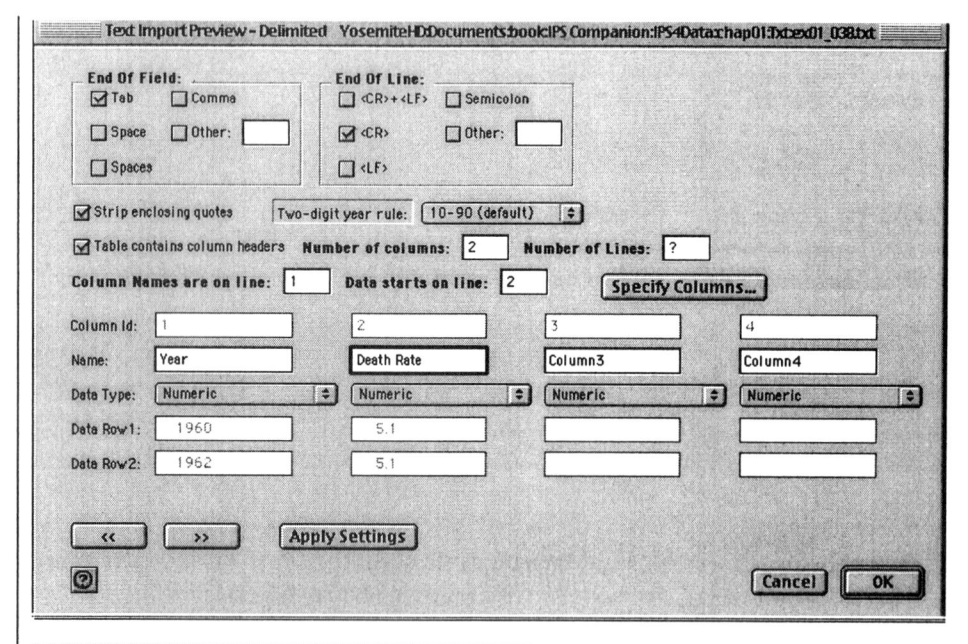

	Year	Death Rate
1	1960	5.1
2	1962	5.1
3	1964	5.4
4	1966	5.5
5	1968	5.2
6	1970	4.7
7	1972	4.3
8	1974	3.5
9	1976	3.2
10	1978	3.3
11	1980	3.3
12	1982	2.8
13	1984	2.6

To look for patterns in the annual motor vehicle death rates, use the **Chart** platform on the **Graph** menu.

1. Select **Graph ⇨ Chart**.

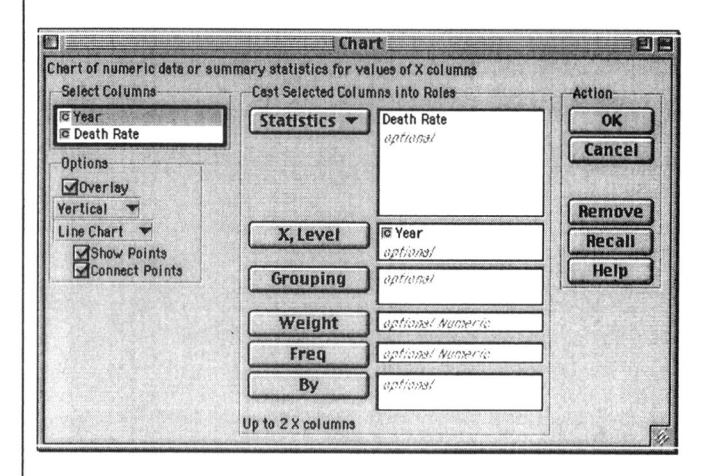

2. Select **Death Rate** from the list of columns.

3. Select **Statistics** ⇨ **Data**.

4. Select **Year** from the list of columns and click **X, Level**.

5. Select **Line Chart** from the pull-down menu titled **Bar Chart** under **Options**.

6. Click **OK**.

7. Click on the red triangle in the **Chart** title bar and deselect **Show Level Legend** to remove the legend for the X-axis.

8. Resize the plot and modify the **Year** axis as in the last example.

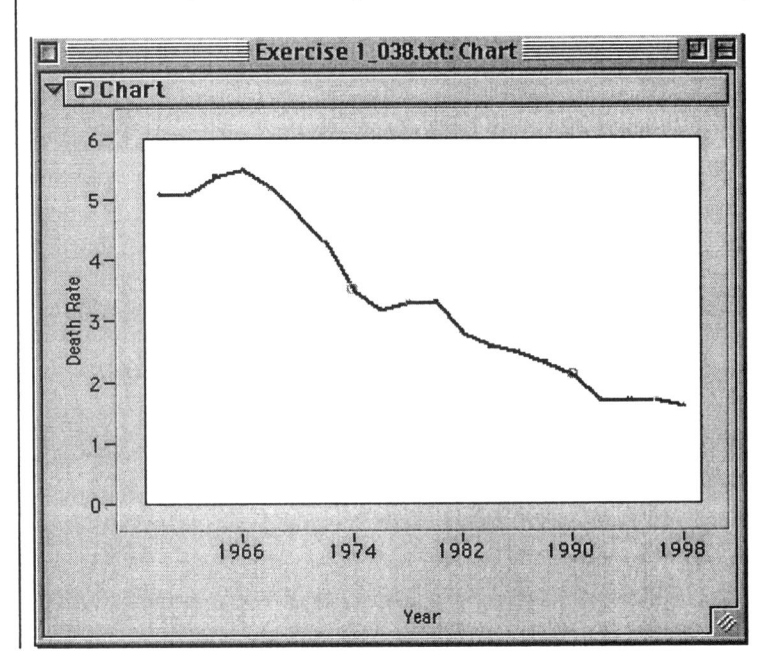

1.2 Describing Distributions with Numbers

All of the numeric summaries presented in IPS are found in the text reports of the *Distribution* platform of JMP.

IPS Example 1.5 Supermarket spending

Example 1.5 of IPS gives the amounts spent by 50 consecutive shoppers in a supermarket. In Section 1.1.2, you displayed the distribution of the variable **Amount Spent**. Let's use the *Distribution* platform of JMP to calculate numerical measures of location and spread for the **Amount spent**.

Recall that the JMP data table **eg01_005.jmp** in the Sample Data folder contains the 50 amounts spent.

1. Select **File** ⇨ **Open** in the menu bar to open the data table.

2. Select the file <u>eg01_005.jmp</u> and press **Open**.

3. Select **Analyze** ⇨ **Distribution**.

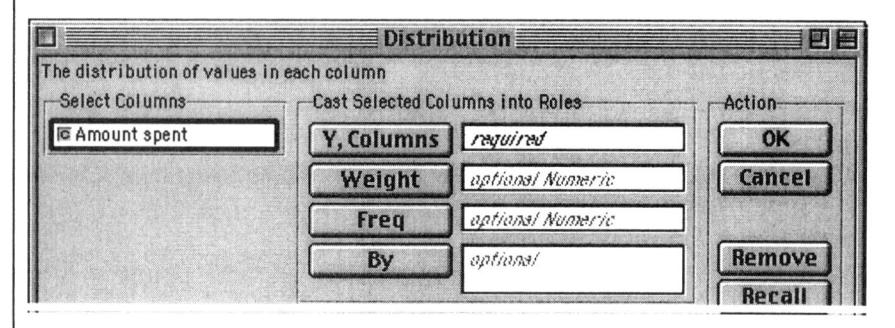

4. Select the column **Amount spent** and press **Y, Columns** and **OK**.

The five-number summary is found in the **Quantiles** report while the mean and standard deviation are found in the **Moments** report. Compare the results with those given in Figure 1.14 of the textbook.

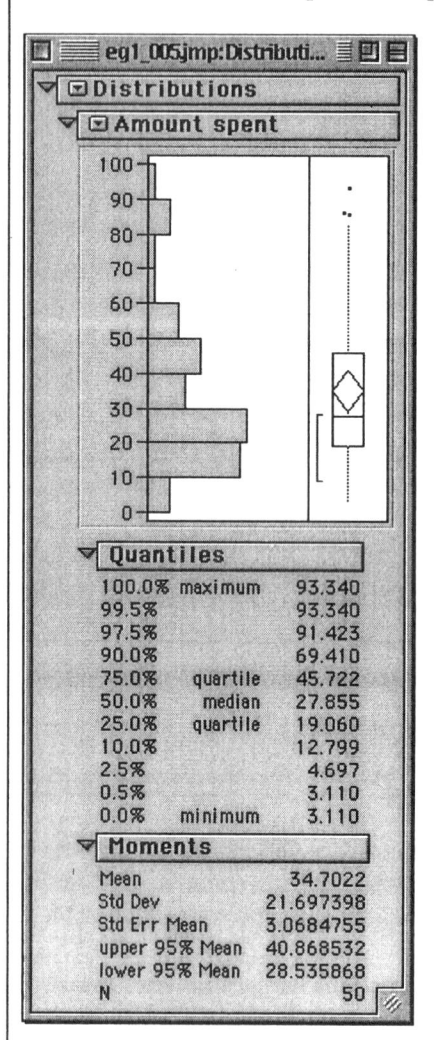

Quantiles

100.0%	maximum	93.340
99.5%		93.340
97.5%		91.423
90.0%		69.410
75.0%	quartile	45.722
50.0%	median	27.855
25.0%	quartile	19.060
10.0%		12.799
2.5%		4.697
0.5%		3.110
0.0%	minimum	3.110

Moments

Mean	34.7022
Std Dev	21.697398
Std Err Mean	3.0684755
upper 95% Mean	40.868532
lower 95% Mean	28.535868
N	50

To display the variance,

1. Click on the red triangle in the **Amount Spent** title bar.

2. Select **Display Options** ⇨ **More Moments** from the menu that opens.

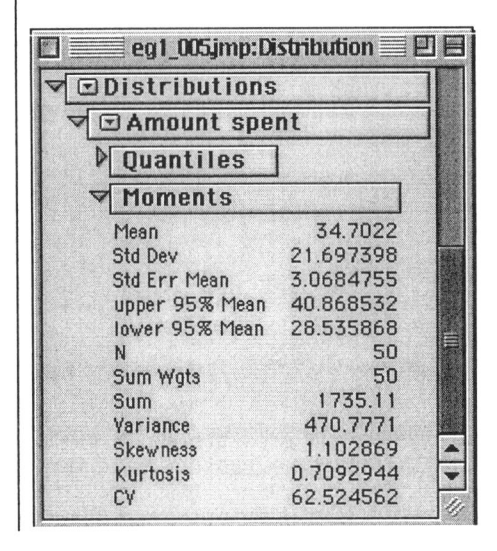

```
eg1_005jmp:Distribution
▽ Distributions
  ▽ Amount spent
    ▷ Quantiles
    ▽ Moments
       Mean           34.7022
       Std Dev        21.697398
       Std Err Mean    3.0684755
       upper 95% Mean 40.868532
       lower 95% Mean 28.535868
       N                  50
       Sum Wgts           50
       Sum             1735.11
       Variance       470.7771
       Skewness         1.102869
       Kurtosis         0.7092944
       CV              62.524562
```

Boxplots

Notice that JMP has automatically displayed a modified boxplot of **Amount spent** and placed it to the right of the histogram. A regular boxplot can be displayed by selecting the **Quantile Box Plot** command in the red triangle menu.

Remark

Some of the other commands in the red triangle menu will be useful in later chapters on statistical inference.

1.3 The Normal Distributions

1.3.1 Normal Distribution Calculations

You can use JMP to calculate areas under normal distributions and find percentiles of a normal distribution. Just put the values in one column and create another column of areas or percentiles using the appropriate normal function.

Example 1.27 NCAA eligibility

The Problem: In 2000, the scores of the more than one million students taking the SATs were approximately normal with mean 1019 and standard deviation 209. The NCAA considers a student a "partial qualifier" eligible to practice and receive an athletic scholarship, but not to compete, if the combined SAT score is at least 720 but not above 820. What percent of all students who take the SAT would be partial qualifiers?

The Solution: In the following steps, you will create a new JMP data table with two columns—**SAT Score** and **Proportion Below**. Then, you will enter 820 and 720 in the first two rows of the first column. Finally, you will use the formula editor to create the corresponding areas under the normal curve to the left of these values.

1. Select **File** ⇨ **New**.

2. Name the first column **SAT Score**.

3. Select **Rows** ⇨ **Add Rows**.

4. Type **2** as the number of rows to add and press **OK**.

5. Type **820** in the first row and **720** in the second.

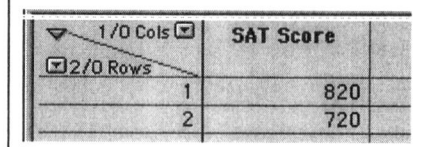

6. Select **Cols** ⇨ **New Column…**

7. Name the column **Proportion Below** and select a format of **Fixed Dec** with **4** decimal places.

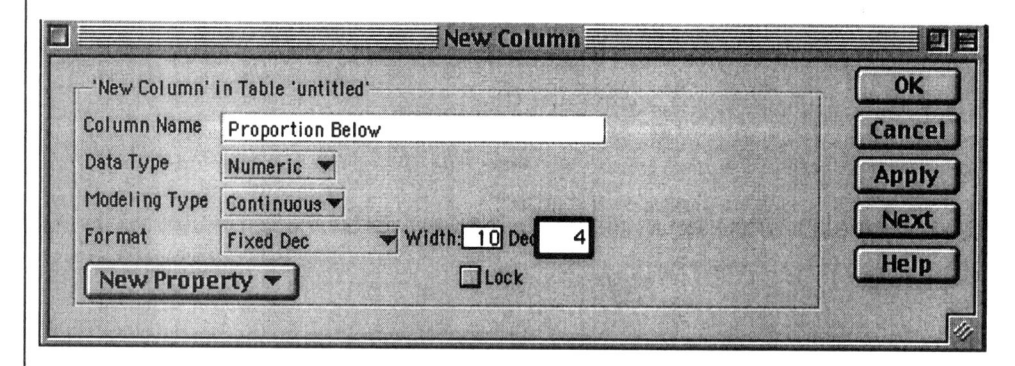

8. Select **New Property** ⇨ **Formula**.

9. In the formula editor, select **Probability** ⇨ **Normal Distribution** from the list of functions.

10. Select **SAT Score** from the list of columns.

11. Press the insert key ^ on the formula keypad, type **1019**, press the insert key ^ again, and then type **209** to specify the normal distribution with mean 1019 and standard deviation 209.

Looking at Data—Distributions

```
┌─ Proportion Below ─────────────────┐
│ Table Columns ▼   ▼  Functions (grouped) ▼   [ OK ]     │
│ SAT Score       [+][-][^]  Row                [ Cancel ]│
│ Proportion Below[×][÷][ℛ]  Numeric                      │
│                 [xʸ][ʸ√x][S] Transcendental  [ Apply ]  │
│                 [±][t=][⊙] Trigonometric                │
│                            Character                    │
│                            Comparison                   │
│                            Conditional      [ Clear ]   │
│                            Probability                  │
│                            Statistical      [ Help ]    │
│                                                         │
│                                                         │
│                                                         │
│         Normal Distribution[ SAT Score , 1019 , 209 ]   │
│                                                         │
│                                                         │
└─────────────────────────────────────────────────────────┘
```

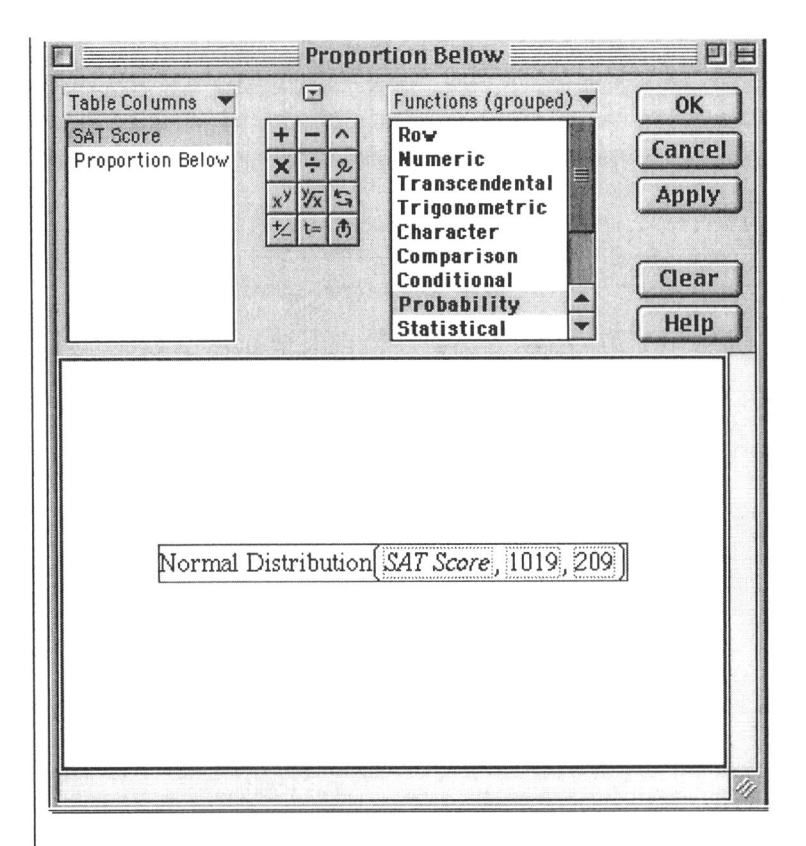

12. Press **OK** and **OK** again in the New Column dialog.

From the resulting data table, we see that the proportion of students with SAT scores in the interval $720 \le X \le 820$ is then $0.1705 - 0.0763 = .0942$. The slight difference from the textbook occurs because the software does not round the corresponding standardized (Z) values to 2 decimal places.

	SAT Score	Proportion Below
1	820	0.1705
2	720	0.0763

Suppose that you want to do the reverse. Find the observed value corresponding to a given relative frequency. This requires the use of the inverse function *normal quantile*.

Example 1.28 Finding percentiles of normal distributions

Scores on the SAT verbal test in recent years follow approximately the $N(505, 110)$ distribution. How high must a student score in order to place in the top 10% of all students taking the SAT?

You reverse the process this time. Create a column with the proportion or area below the X value and use a formula to calculate the corresponding X value.

1. Select **File** ⇨ **New**.

2. Name the first column **Proportion Below**.

3. Type **.90** in the first row and first column of the data table.

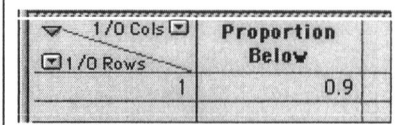

4. Select **Cols** ⇨ **New Column...**

5. Name the column **SAT Score** and select a format of **Fixed Dec** with **1** decimal place.

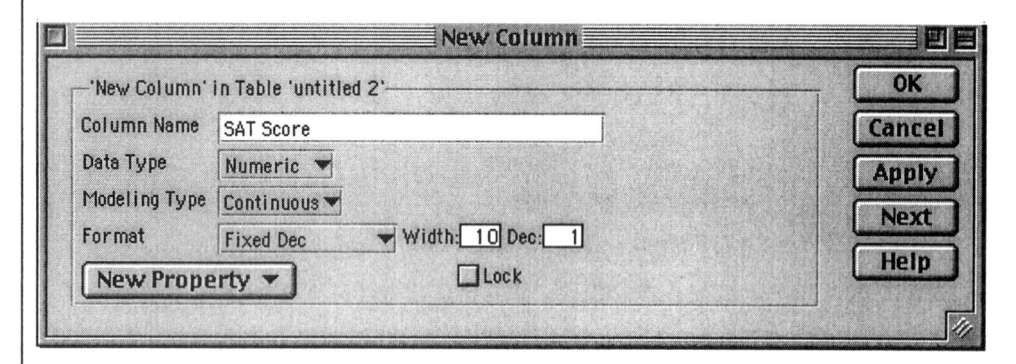

6. Select **New Property** ⇨ **Formula**.

7. In the formula editor, select **Probability** ⇨ **Normal Quantile** from the Function Browser.

8. Select **Proportion Below** from the list of columns.

9. Press the insert key ^ on the formula keypad, type **505**, press the insert key ^ again, and then type **110** to specify the normal distribution with mean 505 and standard deviation 110.

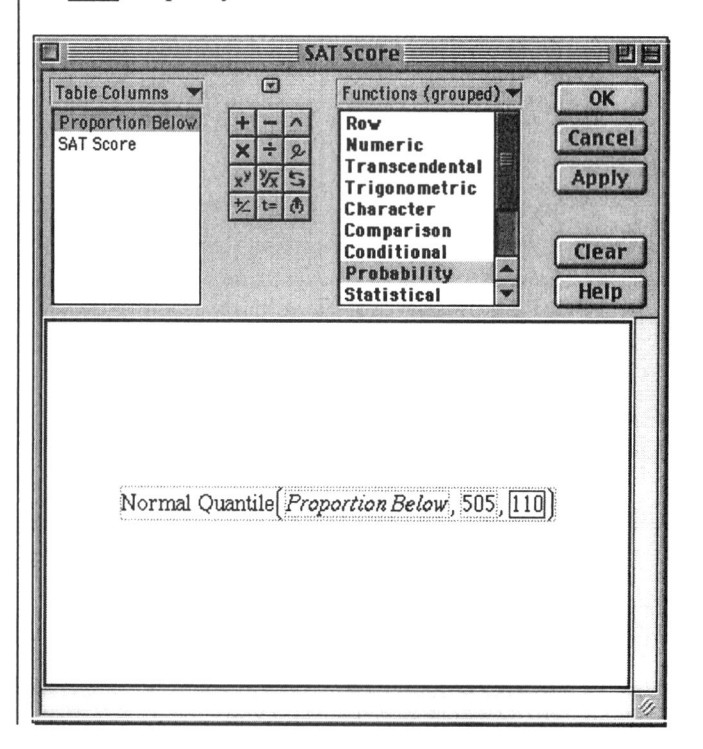

10. Press **OK** and **OK** again in the **New Column** dialog.

	Proportion Below	SAT Score
1	0.9	646.0

From the table, we see that a student must score at least 646 to be in the top 10%.

1.3.2 Assessing Normality: Normal Quantile Plots

Normal quantile plots are easily constructed in JMP using the **Normal Quantile Plot** command in the **Distribution** analysis platform.

Example 1.29 Newcomb's passage time of light

1. Open the JMP data table **ta01_001.jmp** that you saved in Section 1.1.3.

2. Select **Analyze** ⇨ **Distribution** from the menu bar.

3. Select **Passage time** and press **Y, Columns** and **OK** in the dialog window.

All 66 measurements are included and the distribution is skewed to the left by the two outliers.

To get a normal quantile plot for Newcomb's data,

4. Press the red triangle in the **Passage time** title bar and select **Normal Quantile Plot**.

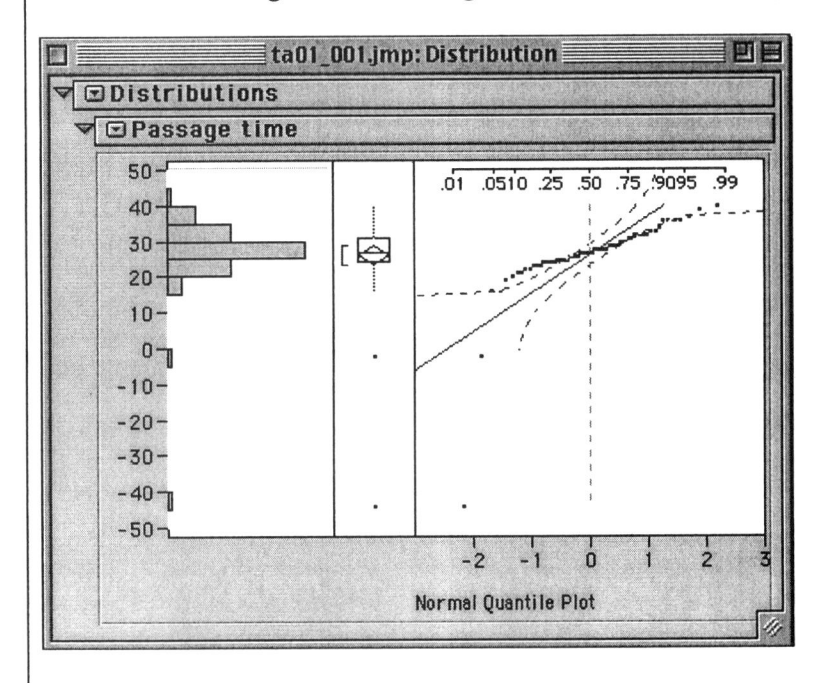

Notice that most of the points in the normal quantile plot nearly form a straight line. The two outliers deviate from that line substantially. JMP enhances the normal quantile plot with reference bands and a reference line. In general, if points fall outside the reference bands, it's not a good idea to use the normal model to describe the distribution of a variable.

Let's exclude the two outliers and see if the normal curve can be used to model the remaining data. Move the JMP data table **ta01_001.jmp** window to the front using the **Window** menu.

1. Click on row **6**. Select **Exclude/Unexclude** from the **Rows** menu.

2. Click on row **10**. Select **Exclude/Unexclude** from the **Rows** menu.

	Passage Time
1	28
2	26
3	33
4	24
5	34
⊘ 6	-44
7	27
8	16
9	40
⊘ 10	-2
11	29
12	22
13	24

3. Bring the **ta01_001.jmp: Distribution** window forward.

4. Click on the red triangle in the **Distributions** title bar.

5. Select **Script** ⇨ **Redo Analysis**. JMP remembers what you did!

Examine the new normal quantile plot. No points lie outside the bands now; the normal curve can be used to describe the remaining data. Compare the normal quantile plots with those in Figures 1.31 and 1.32 of your textbook.

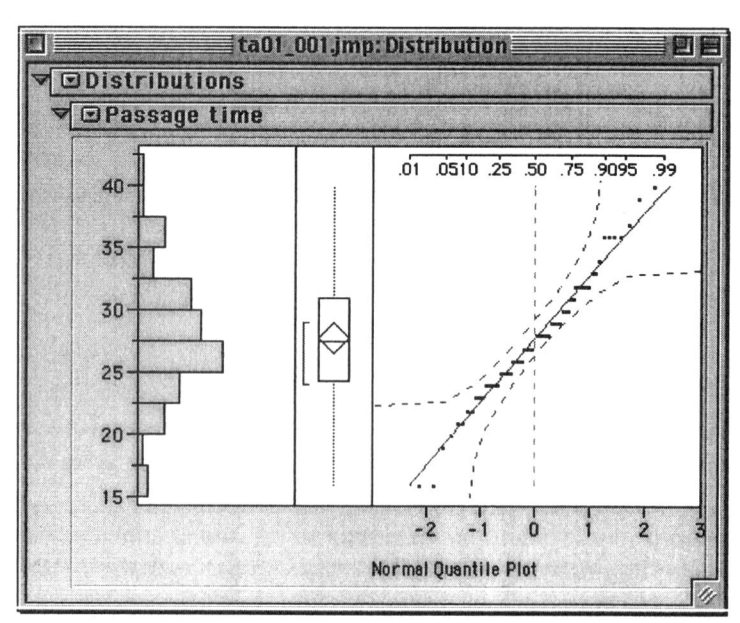

1.4 Summary

With one exception, all statistical graphs and computations in this chapter are performed using the **Distribution** command in the **Analyze** menu. The exception is a time plot. In JMP INTRO, use the **Chart** command in the **Graph** menu for a time plot. In JMP IN and JMP, use the **Time Series** command in the **Analyze** platform.

1.5 Exercises

1. Create a JMP data table similar to the data table **fg01_001.jmp** containing the data on undergraduate students in Exercise 1.11 of IPS. Use JMP to present the data in a bar chart.

Use JMP to help carry out the following exercises from the textbook. Data for the exercises can be imported from text files on the IPS CD-ROM with names corresponding to the exercise number or associated table number and a suffix of *.txt*.

2. Reading test scores. Exercise 1.26.

3. Cavendish's density of earth measurements. Exercise 1.27.

4. Survival times. Exercise 1.30.

5. Treasury bill rates. Exercise 1.37.

6. Winning times of the Boston Marathon. Exercise 1.40. Try the following:
    a. Assign a different color to males and females using the command **Color or mark by Column...** in the **Rows** menu. (Note that **Gender** is the first column.)
    b. Use the **Fit Y by X** platform in the **Analyze** menu. Let **Time** be the **Response** and **Year** be the **Factor**.

7. College costs. Exercise 1.41.

8. SSHA scores. Exercise 1.50. Use the **Subset** command in the **Tables** menu to create a data table of only the women's scores. Don't delete the outlier from the data table to exclude it. Change its row state to exclude it. (See Chapter 0.)

9. Duration of pregnancies. Exercise 1.99.

10. Cavendish's density of earth measurements. Exercise 1.111.

11. Self-concept scores for children. Exercise 1.113

Chapter 2

Looking at Data—Relationships

This chapter studies relationships between variables. Following an approach similar to Chapter 1, relationships are displayed with graphs; the strength of a linear relationship is described by a number; and then straight-lines are used as models for relationships between two quantitative variables. We also learn how to turn complex relationships into linear ones. Discussion of relations between categorical variables is postponed until Chapter 9.

All graphs and statistical computations in this chapter are performed in the second platform **Fit Y by X** of the **Analyze** menu.

2.1 Displaying Relationships with Graphs

A *scatterplot* displays the relationship between two quantitative variables. *Side-by-side boxplots* and *side-by-side means diamonds* display the relationship between a categorical explanatory variable and a quantitative response variable. In JMP, you specify the role (response or categorical) and modeling type of each variable and JMP automatically performs the appropriate methodology.

2.1.1 Two Quantitative Variables: Scatterplots

Scatterplots are created whenever the **Fit Y by X** platform is called and both variables are quantitative.

IPS Figure 2.1 State SAT scores

Figures 2.1 and 2.2 of the textbook show scatterplots to investigate the relationship between a state's mean SAT mathematics score and the percent of its high school seniors who take the exam. The JMP data table **fg02_001.jmp** in the Sample Data folder contains education and related data for the states.

To create a scatterplot,

1. Select **File** ⇨ **Open** and the file **fg02_001.jmp** located in the Sample Data folder.

	State	Region	Pop	SAT Verbal	SAT Math	Percent taking	Percent no HS	Teacher's pay
1	AL	ESC	4273	565	558	8	33.1	31.3
2	AK	PAC	607	521	513	47	13.4	49.6
3	AZ	MTN	4428	525	521	28	21.3	32.5
4	AR	WSC	2510	566	550	6	33.7	29.3
5	CA	PAC	31878	495	511	45	23.8	43.1
6	CO	MTN	3823	536	538	30	15.6	35.4
7	CT	NE	3274	507	504	79	20.8	50.3
8	DE	SA	725	508	495	66	22.5	40.5
9	DC	SA	543	489	473	50	26.9	43.7
10	FL	SA	14400	498	496	48	25.6	33.3

2. Select **Analyze** ⇨ **Fit Y by X**.

Since **SAT Math** is the *response variable* and **Percent taking** is the *explanatory variable*,

3. Select the column **SAT Math** and click **Y, Response**.

4. Select the column **Percent taking** and click **X, Factor**.

5. Press **OK**.

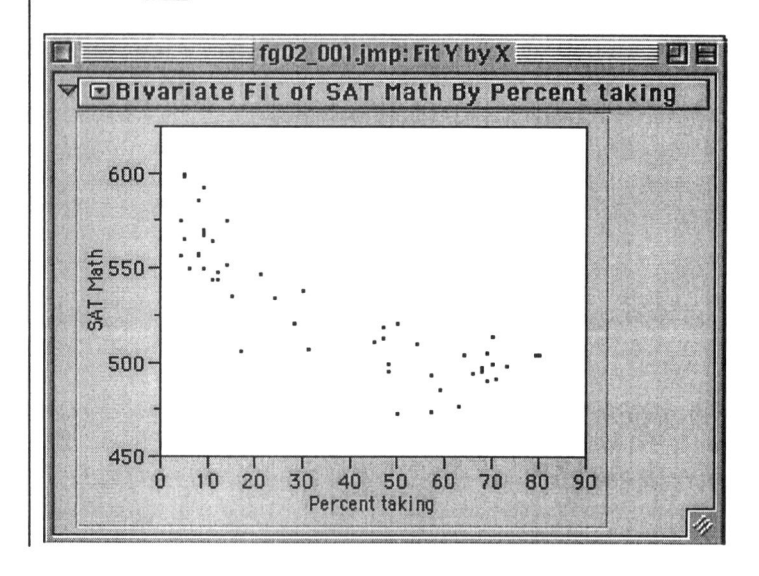

Identifying Individuals on the Scatterplot

If you hover the cursor over a data point, the row number of the state that the point represents is displayed. The **Label/Unlabel** command in the **Cols** menu tells JMP to use a column's values to identify points in plots.

IPS Figure 2.1 State SAT scores (cont'd.)

1. Select the column **State** in the data table.

2. Select **Cols** ⇨ **Label/Unlabel**.

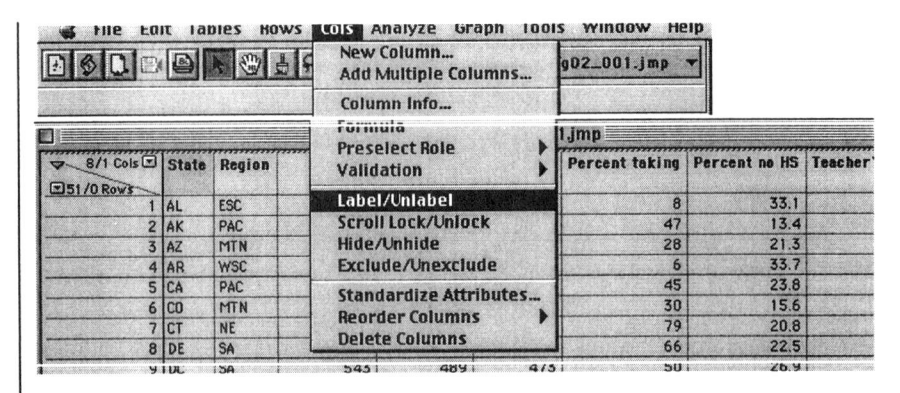

Return to the **fg02_001.jmp: Fit Y by X** window directly or by using the **Window** menu. Move the cursor over the data points. JMP now displays the value of the variable **State**.

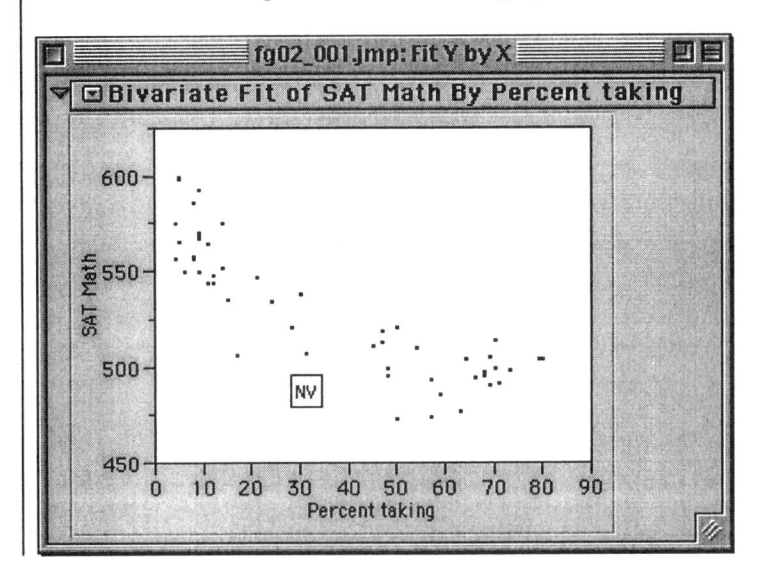

Remarks

The scatterplot can be enhanced in several ways (see Section 0.6.2 in Chapter 0 for details):

- Increase (or decrease) the size of the plot by selecting a corner of the plot and dragging.
- Scroll either axis by moving the hand tool over the numbers.
- Modify tick marks and the increment between numbers by double-clicking on a scale data value.
- Modify or enhance an axis name by double-clicking on the axis name.
- Create editable notes to be displayed and stored with the plot using [A], the annotate tool.

Adding Categorical Variables to Scatterplots

IPS Figure 2.1 State SAT scores (cont'd)

The Census Bureau groups the states into regions of the country. To investigate regional patterns, we might wish to assign the points on the scatterplot associated with the individual states different colors and symbols depending on which region they are in. To do this, we change the *state* of the rows/individuals. (See Section 0.3.2 in Chapter 0 for more details on row states.) The *row state characteristic* that we use here is called **Color or Marker by Column**. We wish to color and mark the points differently depending on the values of a variable or column.

1. Select Rows ⇨ Color or Marker by Column.

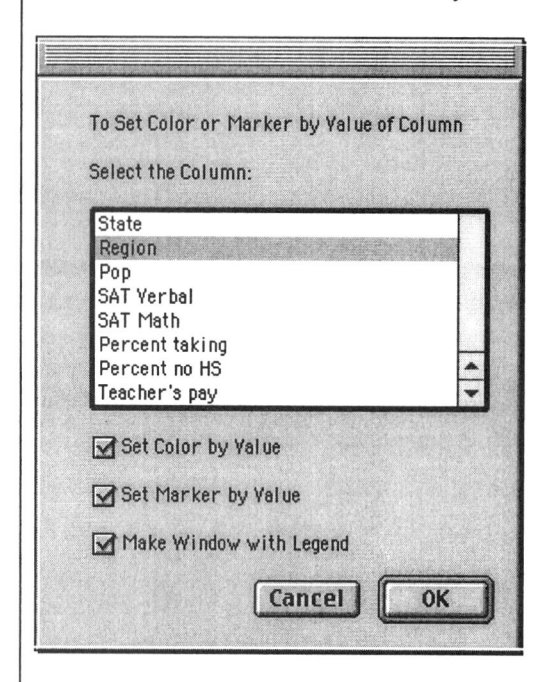

2. Select Region, and check Set Marker by Value and Make Window with Legend.

3. Press OK.

Return to the **fg02_001: Fit Y by X** window and compare two regions.

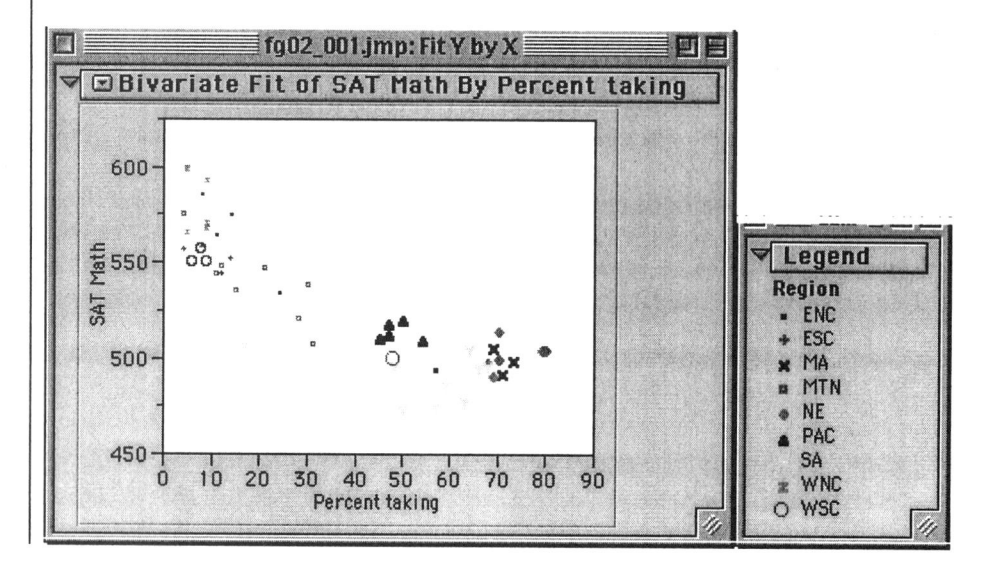

2.1.2 A Categorical Explanatory Variable and a Quantitative Response

Variable: Side-by-Side Boxplots and Means Diamonds

To display a relationship between a categorical explanatory variable and a quantitative response variable, we make a side-by-side comparison of the distributions of the response for each category. Some tools for such comparisons are

- Side-by-side boxplots. See IPS Figure 1.16 on the comparison of calories (the quantitative response) for beef, meat, and poultry hot dogs (the categories).

- Side-by-side point plots.

- Side-by-side means diamonds.

All three plots are available in JMP. We will use the hot dog data to illustrate.

IPS Table 1.9 Differences among types of hot dogs

People who are concerned about health may prefer low-calorie hot dogs and ask "Are there any systematic differences among the three types of hot dogs in calories?" In other words, "Are the type and calorie content of a hot dog related?"

The JMP data table **Hot Dogs** located in the Sample Data folder contains data on 54 brands of hot dogs.

1. Select **File** ⇨ **Open** in the menu bar to open the data table.

2. Select the file **Hot Dogs.jmp** located in the Sample Data folder.

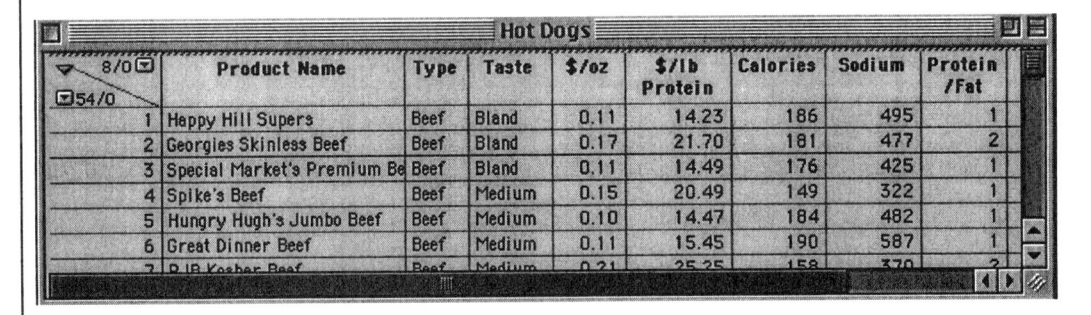

Product Name	Type	Taste	$/oz	$/lb Protein	Calories	Sodium	Protein/Fat
1 Happy Hill Supers	Beef	Bland	0.11	14.23	186	495	1
2 Georgies Skinless Beef	Beef	Bland	0.17	21.70	181	477	2
3 Special Market's Premium Be	Beef	Bland	0.11	14.49	176	425	1
4 Spike's Beef	Beef	Medium	0.15	20.49	149	322	1
5 Hungry Hugh's Jumbo Beef	Beef	Medium	0.10	14.47	184	482	1
6 Great Dinner Beef	Beef	Medium	0.11	15.45	190	587	1

There are 54 brands of hot dogs and 8 variables describing each brand. First, examine the distribution of each of the variables **Type** and **Calories**.

3. Select **Analyze** ⇨ **Distribution**.

4. Select the columns **Type** and **Calories** and press **Y, Columns** and **OK**.

Hot Dogs: Distribution

Distributions

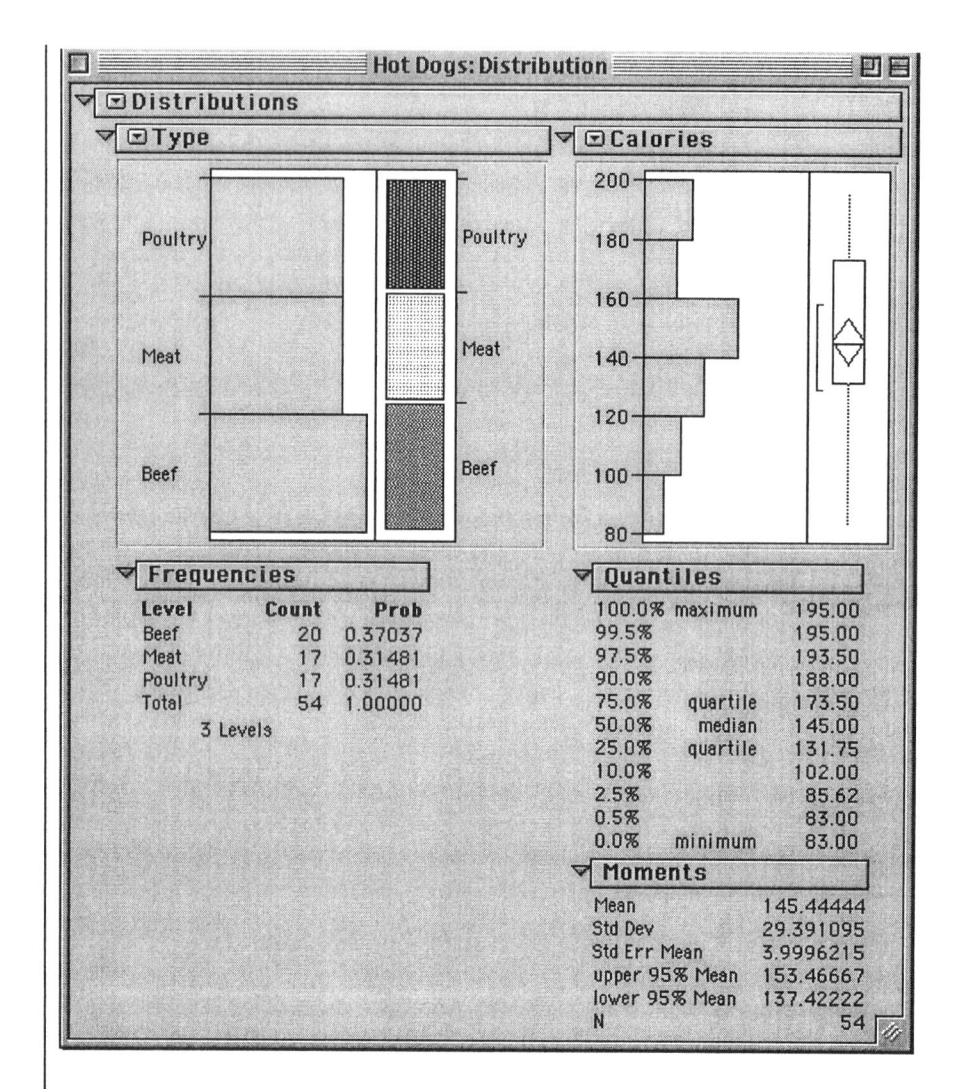

Type

Calories

Frequencies

Level	Count	Prob
Beef	20	0.37037
Meat	17	0.31481
Poultry	17	0.31481
Total	54	1.00000

3 Levels

Quantiles

100.0%	maximum	195.00
99.5%		195.00
97.5%		193.50
90.0%		188.00
75.0%	quartile	173.50
50.0%	median	145.00
25.0%	quartile	131.75
10.0%		102.00
2.5%		85.62
0.5%		83.00
0.0%	minimum	83.00

Moments

Mean	145.44444
Std Dev	29.391095
Std Err Mean	3.9996215
upper 95% Mean	153.46667
lower 95% Mean	137.42222
N	54

Remark

Notice that the calories for all three types of hot dogs are listed in one column of the JMP data table and not three, even though Table 1.9 of IPS uses three columns, one for each type of hot dog. This is because each row of a JMP data table represents an individual; in this case, it is a brand of hot dog. Since each column in a JMP data table represents a variable, the three columns of Table 1.9 in IPS are stacked into two variables—**Type** and **Calories**. This is very important to remember since all statistical computations and graphs assume that the individuals are the rows of a data table and the variables are columns.

To make a side-by-side comparison of the distributions of the calories for each type of hot dog, use the **Fit Y by X** platform.

5. Select <u>Analyze</u> ⇨ <u>Fit Y by X</u>.

Since **Calories** is the *response variable* and **Type** is the *explanatory variable*,

6. Select the column <u>Calories</u> and click <u>Y, Response</u>.

7. Select the column <u>Type</u> and click <u>X, Factor</u>.

8. Press **OK**.

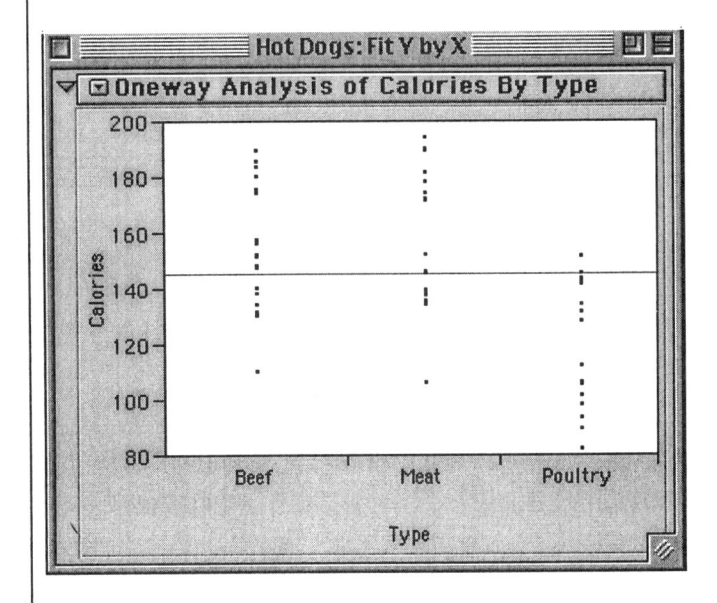

[Screenshot: Hot Dogs: Fit Y by X — Oneway Analysis of Calories By Type showing side-by-side point plots for Beef, Meat, and Poultry with a horizontal line near 145.]

JMP presents *side-by-side point plots*. The horizontal line at 145.4 calories is the overall mean calorie content. To get *side-by-side boxplots*,

9. Press the red triangle in the **Oneway Analysis of Calories By Type** report.

10. Select **Display Options** ⇨ **BoxPlots** from the menu that opens.

11. Deselect **Display Options** ⇨ **Points** to remove the points.

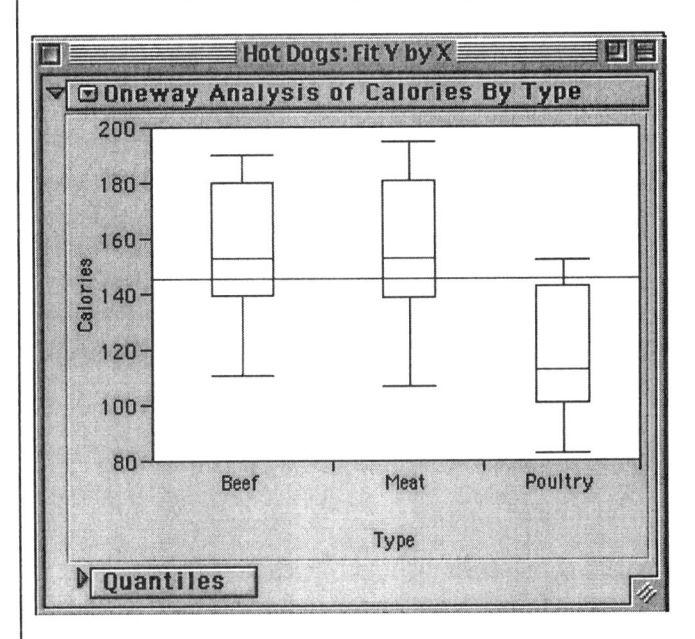

[Screenshot: Hot Dogs: Fit Y by X — Oneway Analysis of Calories By Type showing side-by-side boxplots for Beef, Meat, and Poultry.]

Side-by-side means diamonds are another tool, offered by JMP, for comparing the distributions of a response variable among the categories of another variable.

12. Deselect **Display Options** ⇨ **BoxPlot**.

13. Select **Display Options** ⇨ **Mean Diamonds**.

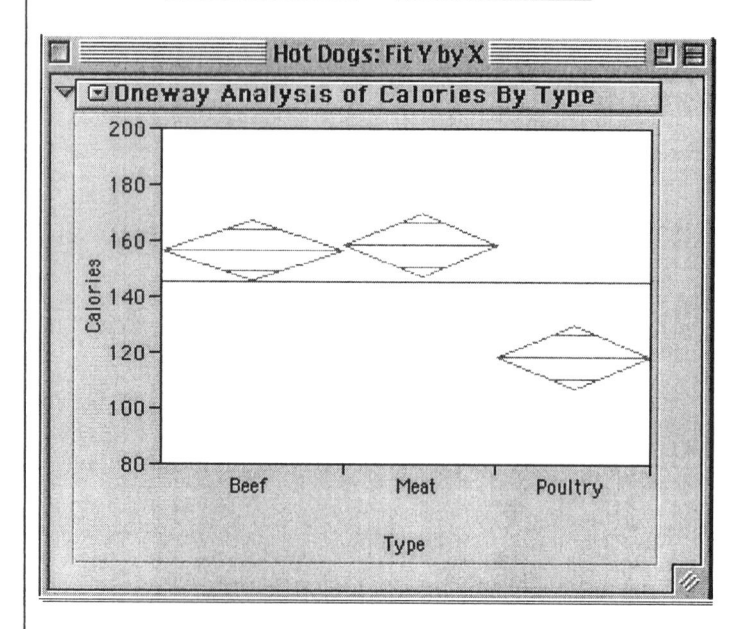

In both displays, it can be seen that Poultry hot dogs have lower calories on average than either Beef or Meat hot dogs. Notice also that there appears to be no difference between Beef and Meat hot dogs.

Remark

The remainder of this chapter concentrates on relationships among quantitative variables. We will use *side-by-side means diamonds* again in conjunction with the analyses discussed in Chapters 7, 12, and 13 of the textbook. To learn more about *means diamonds*, select the [?] tool and click on a diamond.

2.2 Describing Relationships with Numbers: Correlation

To find the *correlation* between two quantitative variables, we use the **Density Ellipse** command in the **Fit Y by X** platform in JMP INTRO.

IPS Figure 2.1   State SAT scores revisited

The scatterplot for a state's mean SAT mathematics score and the percent of its high school seniors who take the exam shows a somewhat strong negative linear relationship between the variables. Let's calculate the *correlation*. We first display the scatterplot.

1. Open the JMP data table **fg02_001.jmp** located in the Sample Data folder if it is closed.

2. Select Analyze ⇨ Fit Y by X.

3. Select the column **SAT Math** and click **Y, Response**.

4. Select the column **Percent taking** and press **X, Factor** and **OK**.

5. Click on the red triangle in the **Bivariate Fit** report title and select Density Ellipse ⇨ .95. (It doesn't matter which number you select.)

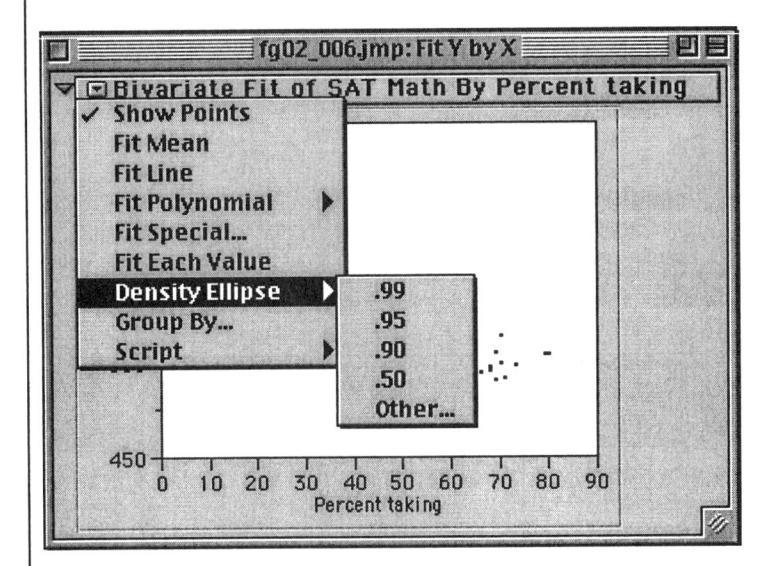

6. Open the **Correlation** report by clicking the disclosure button next to **Correlation**. Notice that $r = -0.86072$.

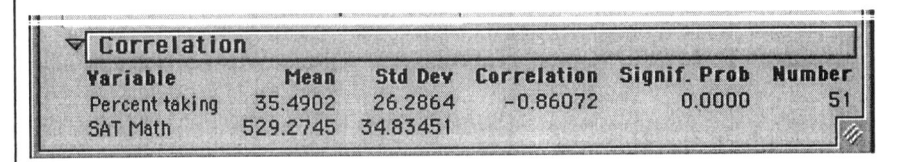

Remark

In JMP IN and in the professional version of JMP, the **Multivariate** analysis platform can also be used to calculate correlations.

2.3 Models for Relationships: Least-Squares Regression

To fit the least-squares regression line, use the **Fit Line** command in the red triangle menu for scatterplots.

IPS Example 2.10 How do children grow?

Because the pattern of growth of children varies from child to child, we can best understand the general pattern by following the average height of a number of children. Table 2.7 of IPS presents the mean heights of a group of children in Kalama, an Egyptian village. We first create a scatterplot to examine the relationship between age and average height. Age is the explanatory variable so we wish to plot it on the x-axis. The text file **ta02_007.txt** on the IPS CD-ROM contains the pairs of values.

1. Import the file **ta02_007.txt** into a JMP data table (see Section 0.2.1 in Chapter 0 for details) and name the columns of data values **Age** and **Height** respectively.

Display a scatterplot of **Height** by **Age**.

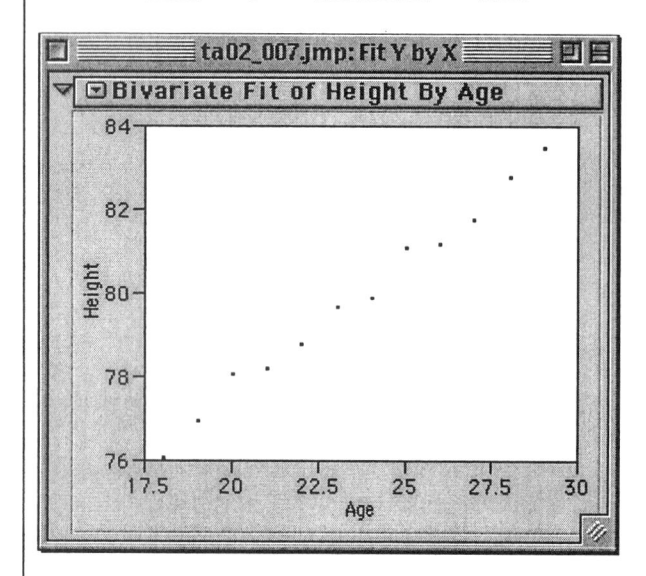

2. Select <u>**Analyze**</u> ⇨ <u>**Fit Y by X**</u>.

3. Select <u>**Height**</u> and <u>**Y, Response**</u>.

4. Select <u>**Age**</u> and press <u>**X, Factor**</u> and <u>**OK**</u>.

The plot shows a strong linear relationship with no outliers. A straight line will serve as a good model for the relationship between the **age** and **height** of Kalama children.

Fit the least-squares regression line to the data.

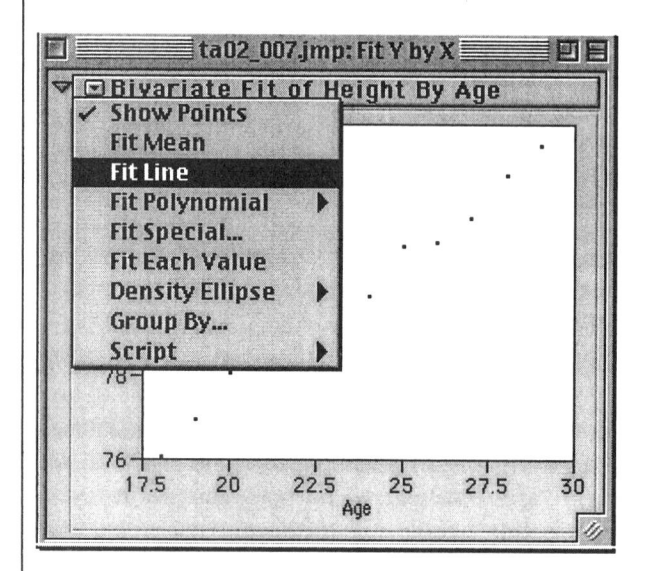

5. Press the red triangle and select <u>**Fit Lin**</u>e from the pop-up menu.

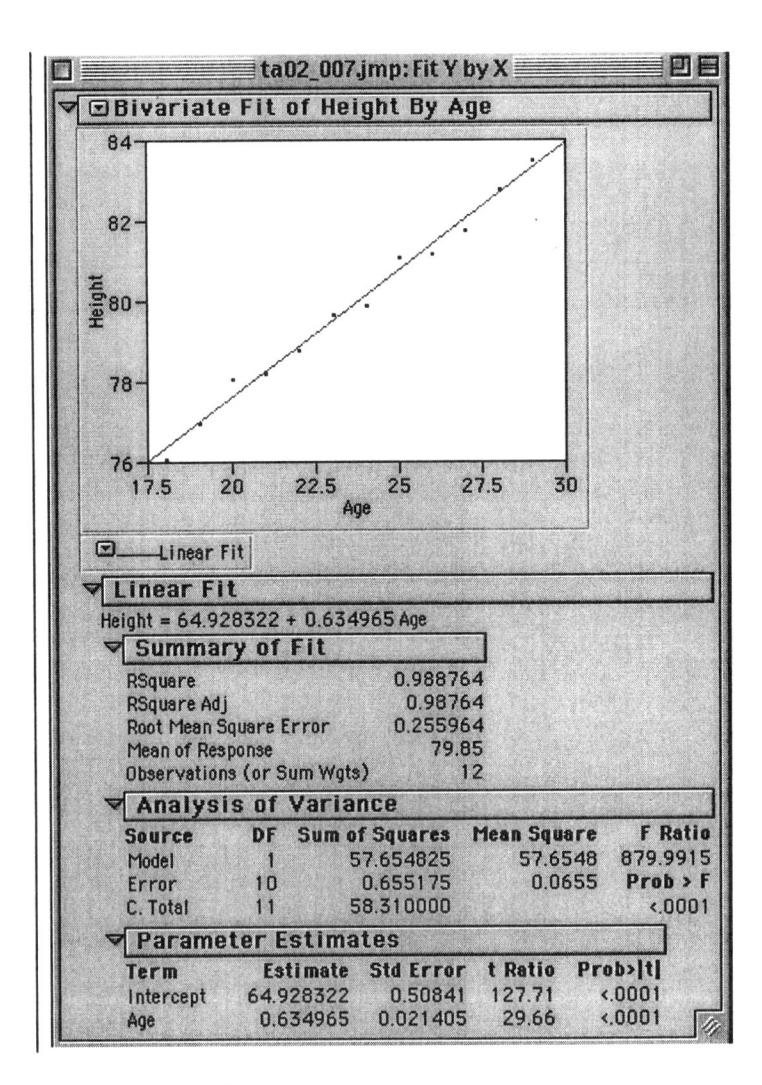

![Bivariate Fit of Height By Age - JMP output showing scatter plot with linear fit, Height = 64.928322 + 0.634965 Age, RSquare = 0.988764, RSquare Adj = 0.98764, Root Mean Square Error = 0.255964, Mean of Response = 79.85, Observations = 12. ANOVA: Model DF 1, SS 57.654825, MS 57.6548, F Ratio 879.9915, Prob > F <.0001; Error DF 10, SS 0.655175, MS 0.0655; C. Total DF 11, SS 58.310000. Parameter Estimates: Intercept 64.928322, Std Error 0.50841, t Ratio 127.71, Prob>|t| <.0001; Age 0.634965, Std Error 0.021405, t Ratio 29.66, Prob>|t| <.0001.]

RSquare, r^2

The least-squares equation can be found directly under the **Linear Fit** title bar **Height = 64.9 + 0.635 Age**; *RSquare (r^2)* can be found directly under the **Summary of Fit** title bar **RSquare = r^2 = 0.988764**. Recall that r^2 is the proportion of the variability in **Height** that is explained by the least squares regression of **Height** on **Age**. To see the variability in **Height** better,

Select Fit Mean from the pop-up menu next to **Bivariate Fit of Height by Age**.

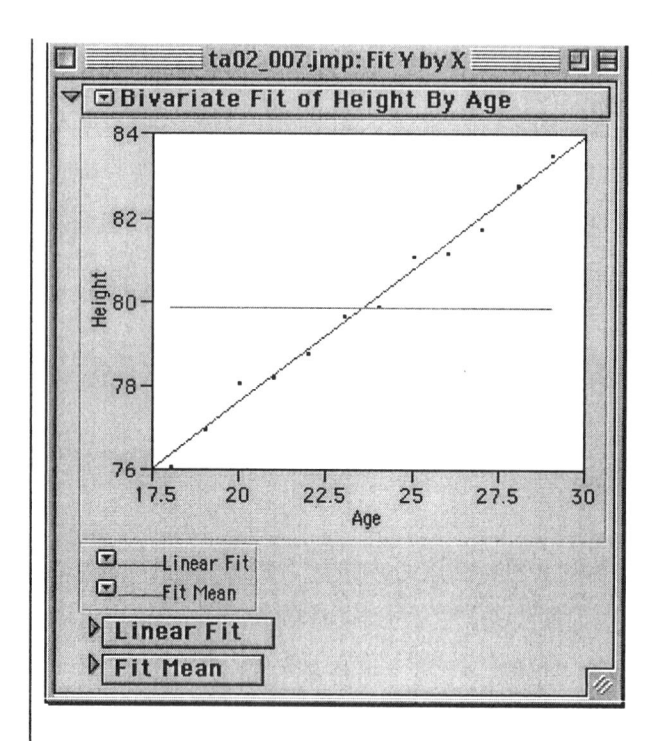

Compare the variability of the points about the horizontal green line with that about the tilted red line. The vertical distances of the points from the red (regression) line are considerably less than from the horizontal green line.

Prediction

We can use JMP to *predict* the response for a specific value of the explanatory variable x. For example, we might want to predict the mean height of Kalama children at 20 months of age.

IPS Example 2.10 How do children grow? (cont'd.)

1. Select the **crosshair** tool from the **Tools** palette.

2. Place the cursor, now resembling a crosshair, on the line directly above 20 and press. The *predicted value* of the mean **Height,** 77.6 inches, for children at 20 months of age is displayed.

JMP can also store the *predicted values* for each of the individuals in the data table.

3. Press the red triangle that is directly below the scatterplot and next to the **Linear Fit** title bar, and select Save Predicteds from the menu that opens.

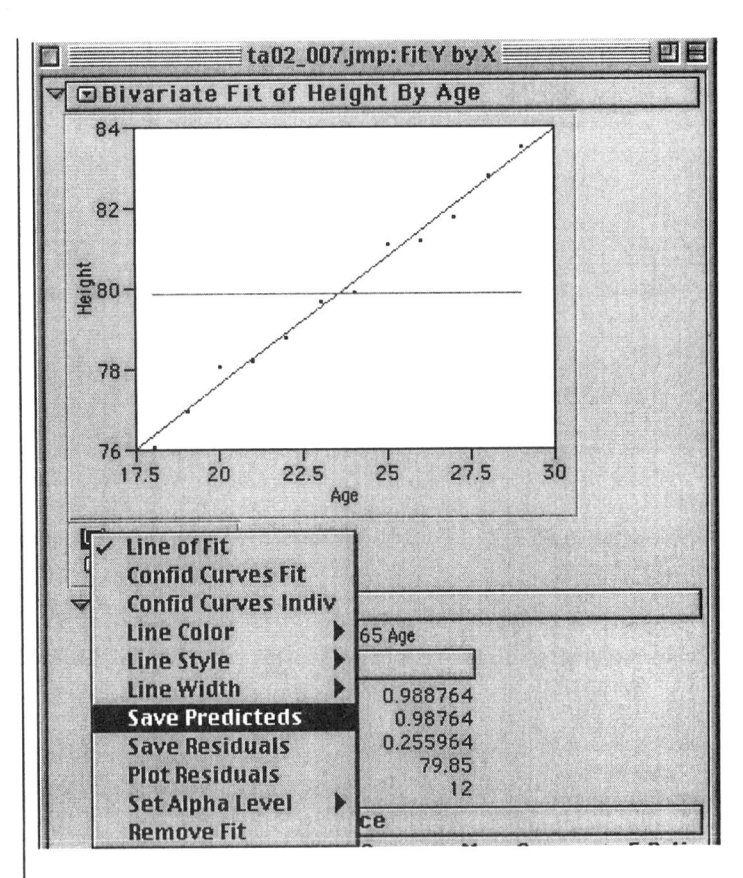

4. Select the **ta02_007** data table window and notice that a new column **Predicted Height** was created to hold the predicted value for each observation.

2.4 Assessing the Fit: Residuals, Outliers, and Influential Observations

Besides fitting models that describe the overall pattern of a relationship, JMP helps to assess the appropriateness of a fitted model and to identify striking deviations from that model. The professional statistician does these tasks first before examining r^2, the equation of the line, or predicting the response for a specific value of the explanatory variable.

Residuals

Residuals are the vertical deviations of the observed data points from the corresponding predicted values on the least-squares regression line. As such, they represent deviations of the regression model from the

76   Looking at Data—Relationships

data points, and a plot of the residuals can help you assess the appropriateness of a regression line as a model for the data. Plotting is a task at which JMP excels. With one command, you can obtain a residual plot; with another command, you can tell JMP to calculate all the residuals and store them in the original data table for later use.

IPS Example 2.10 How do children grow? (cont'd.)

1. Bring the report window ta02_007: Fit Y by X forward. (If you no longer have the window available, repeat the first 5 steps in Section 2.3.)

2. To plot the residuals against the explanatory variable for the linear fit, select Plot Residuals from the pop-up menu located directly below the scatterplot next to **Linear Fit** (not the one next to **Fit Mean**, if you are using the report from Section 2.3).

Since the plot is a random band of points centered at zero, the least-squares model, **Height = 64.9 + 0.635 Age**, provides an appropriate description of the relationship between height and age.

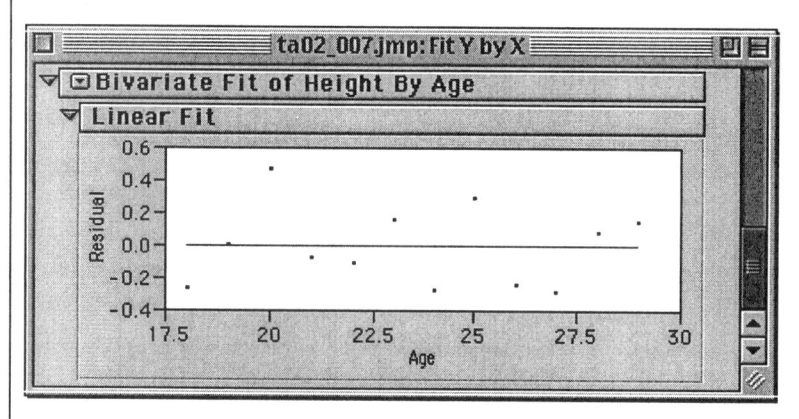

3. To save the residuals to the JMP data table, select Save Residuals from the red triangle menu located directly below the scatterplot next to **Linear Fit**.

Row	Age	Height	Predicted Height	Residuals Height
1	18	76.1	76.3576923	-0.2576923
2	19	77	76.9926573	0.00734266
3	20	78.1	77.6276224	0.47237762
4	21	78.2	78.2625874	-0.0625874
5	22	78.8	78.8975524	-0.0975524

Outliers and Influential Observations

In addition to judging the appropriateness of a regression line as a model for the data, we need to look for striking individual points—*outliers* and *influential observations*. *Outliers* are points that are outlying in the *y*, or vertical, direction while points that are outlying in the x, or horizontal, direction are potentially *influential observations*. Both can be identified using residual plots. To judge the influence of a point outlying in the *x* direction, we must find the regression line with and without the suspect point.

IPS Example 2.18 Age at which a child begins to talk

Example 2.18 of the textbook presents data from a cognitive study of children designed to investigate the relationship of the age at which a child begins to talk and a later score on a test of mental ability. Let's determine if there are any outliers or influential observations in the data. The data file **ta02_009.txt** on the IPS CD-ROM contains the data from this study.

1. Import the text file **ta02_009.txt** on the IPS CD-ROM into a JMP data table (see Section 0.2.1 in Chapter 0 for details) and name the columns **Child**, **Age**, and **Score**.

Fit the least-squares regression line and obtain a residual plot.

2. Select Analyze ⇨ Fit Y by X.

3. Select Score and press Y, Response.

4. Select Age and press X, Factor and OK.

5. Press the red triangle on the **Bivariate Fit of Score by Age** title bar and select Fit Line from the menu that opens.

6. Press the red triangle next to **Linear Fit,** which is directly below the scatterplot, and select Plot Residuals.

7. Identify the children associated with the outlying points by holding the cursor over them.

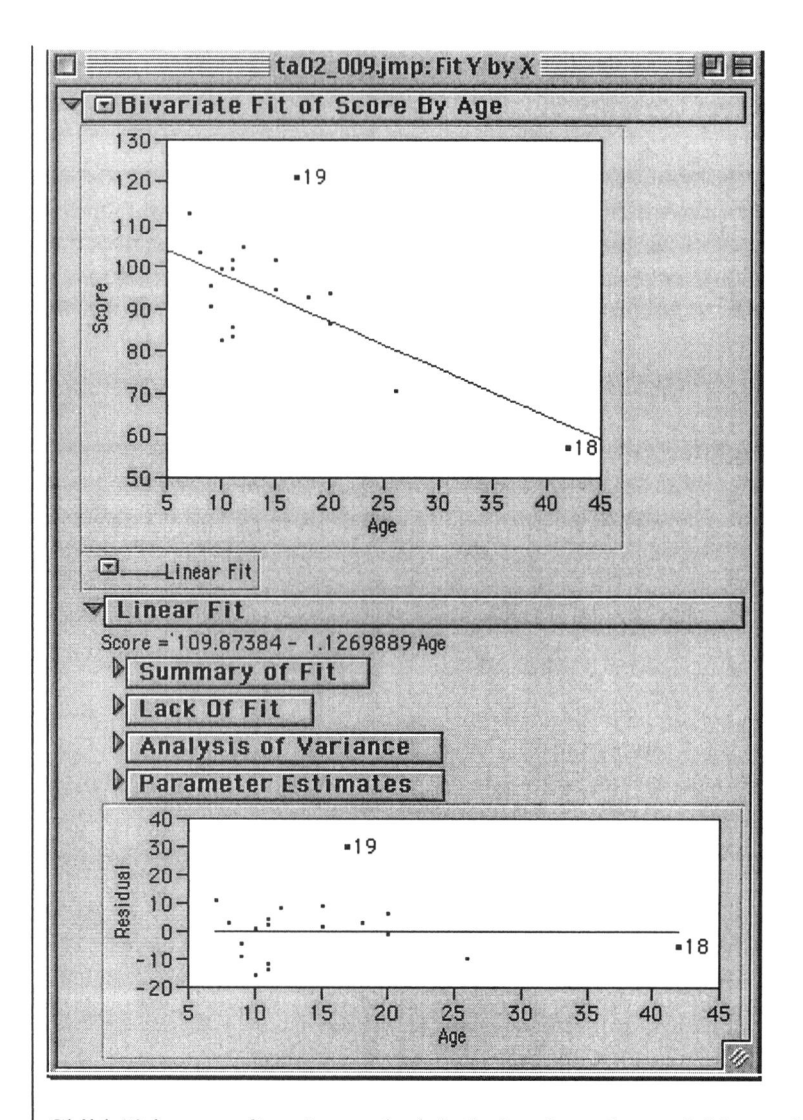

Child 19 is an *outlier*. Its vertical deviation from the model is much larger than for other children. The model does not fit this child well. Child 18, while having a small residual, is outlying in the *x* direction and, as such, is potentially *influential*. Click on it and notice that the corresponding point in the scatterplot of **Score by Age** is highlighted. To investigate the influence of child 18 on the fitted line, you need to exclude this child and refit the least-squares regression line. To do this, simply change the row state of child (row) 18 in the data table to **Exclude**. (Row states are discussed in more detail in Section 0.4 in Chapter 0.)

8. Select Window ⇨ ta02_009 to bring the data table to the front.

9. If row 18 (child 18) is not highlighted, select **Row 18**.

10. Select Rows ⇨ Exclude/Include and notice that the exclusion symbol Ø appears in the row number area next to row 18 at the left of the data grid.

11. Select Window ⇨ ta02_009: Fit Y by X to bring the report to the front.

12. To have JMP automatically duplicate the analysis without child 18, press the red triangle on the **Bivariate Fit of Score by Age** title bar at the top.

13. Select Script ⇨ Redo Analysis.

Compare this scatterplot and the equation of this line with those in the previous report that included child 18. The slope of the least squares regression line has been substantially changed. Child 18 is an influential observation.

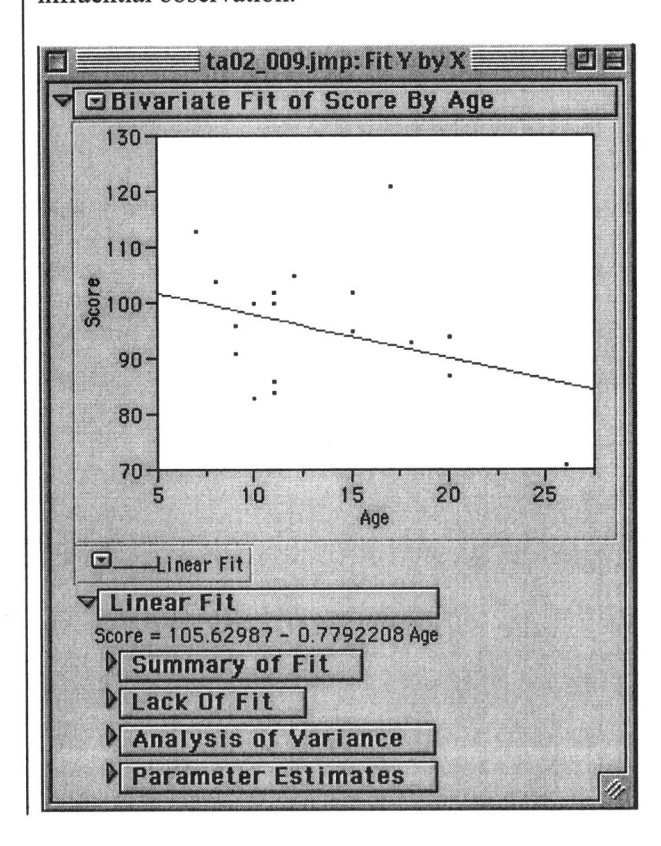

Bivariate Fit of Score By Age

Score = 105.62987 − 0.7792208 Age

2.6 Transforming Relationships

Nonlinear relationships between variables can sometimes be changed into linear ones by transforming one or both of the variables. To transform variables in JMP, you create a new variable using the formula editor.

IPS Exercise 2.92 Moore's law

Gordon Moore, one of the founders of Intel Corporation, predicted that the number of transistors on an integrated circuit chip would double every 18 months. This has become known as "Moore's law" for microprocessing power. The text file **ex02_092.txt** contains data on the number of transistors on microprocessors made by Intel since 1971. Show that a log transformation of the number of transistors indicates linear growth and hence that an exponential growth model is correct.

1. Open the JMP data table **MooresLaw.jmp** that you created in Section 0.2.2 in Chapter 0. If this JMP data table cannot be found, import the text file **ex02_092.txt** on the IPS CD-ROM into a JMP data table and name the columns **Processor**, **Date**, and **Transistors**.

Obtain a scatterplot of the number of transistors versus time.

2. Select Analyze ⇨ Fit Y by X.

3. Select Transistors and press Y, Response.

4. Select Date, and press X, Factor and OK.

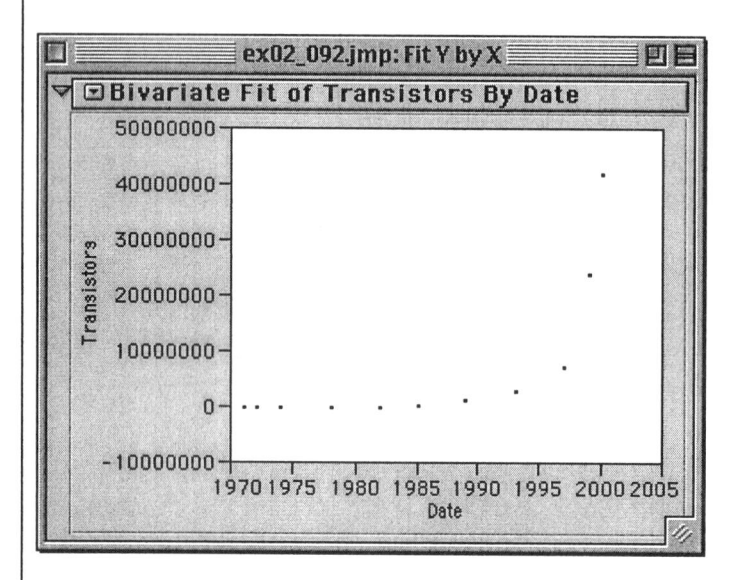

The pattern is characteristic of exponential growth—slow growth for an extended period followed by explosive growth. If it is exponential growth, a logarithmic transformation of the number of transistors will straighten the pattern. We will create a new variable to hold the logarithms of the number of transistors and plot it against time.

5. Go to the JMP data table **ex02_092** and select Cols ⇨ New Column to create a column of logarithms of the values in the column **Transistors**.

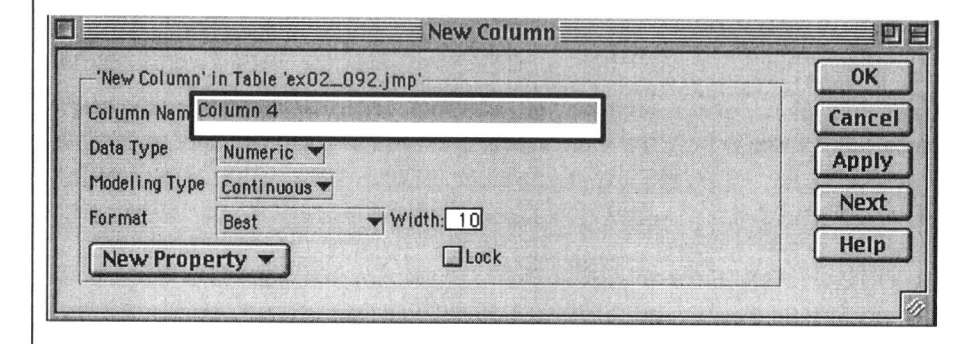

6. Enter **Log Transistors** in the **Column Name** field of the **New Column** panel.

7. Select Fixed Dec from the **Format** menu and enter **3**.

8. Press New Property and select Formula.

9. From the list of columns, select Transistors.

10. From the list of **Functions (grouped)**, select Transcendental ⇨ Log.

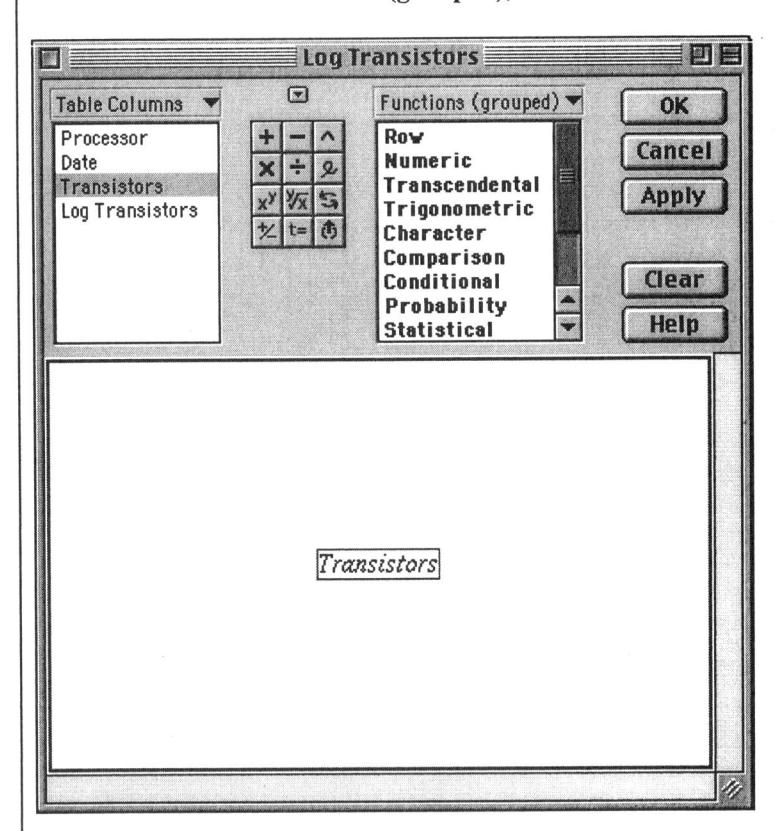

11. Press OK and OK again.

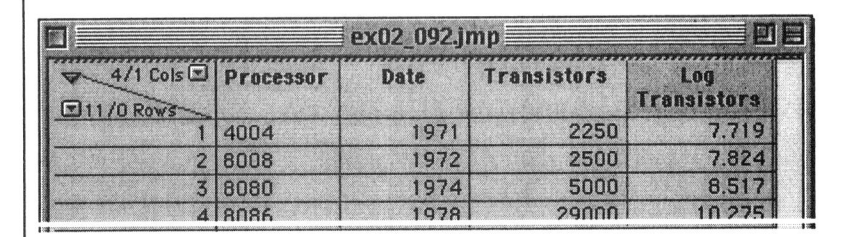

Now plot the logarithm of the number of transistors against time and fit a least-squares regression line to the data.

12. Select Analyze ⇨ Fit Y by X.
    a. Select Log Transistors and press Y, Response.
    b. Select Date, and press X, Factor and OK.

13. Press the red triangle next to the **Bivariate Fit of Log Transistors by Date** title bar and select Fit Line from the menu that opens.

The relationship between the logarithms of the number of transistors and time is quite linear. Hence, exponential growth is a good model for the relationship of the number of transistors and time.

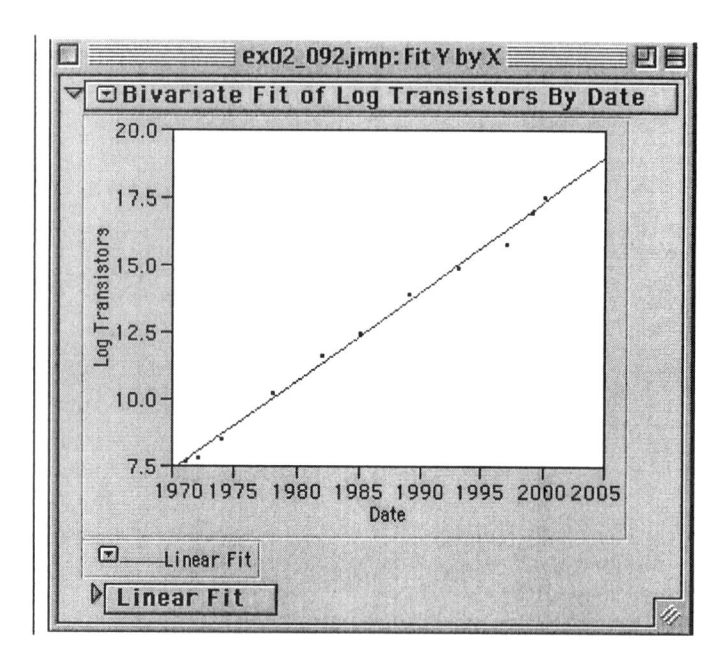

2.7 Summary

All graphs and statistical computations in this chapter are performed in the second platform **Fit Y by X** of the **Analyze** menu.

Graph/Computation	Command
Displaying relationships	Analyze ⇨ Fit Y by X
Scatterplots	
Side-by-side boxplots	
Side-by-side means diamonds	
Describing relationships	Analyze ⇨ Fit Y by X ⇨ Density Ellipse
Models for relationships	Analyze ⇨ Fit Y by X ⇨ Fit Line

To transform a variable, create a new variable using the **New Column** command, on the **Cols** menu, and the formula editor.

2.8 Exercises

Use JMP to help carry out the following exercises from the textbook. Data for the exercises can be imported from text files on the IPS CD-ROM with names corresponding to the exercise number or associated table number and a suffix of .txt.

1. Wine consumption and heart attacks. Exercise 2.7.

2. World record times for 10,000-meter races. Exercise 2.15.

3. Fidelity Investments sector funds. Exercise 2.17.

4. Color attraction for cereal leaf beetles. Exercise 2.18.

5. Vanguard International Growth Fund and the EAFE index. Exercise 2.27.

6. Speed and gas consumption. Exercise 2.29.

7. World record times for 10,000-meter races. Exercise 2.45. The **Group By** command in the **Fitting** menu (next to the scatterplot name) can tell JMP to fit separate regression lines for men and women.

8. Literacy. Exercise 2.55.

9. Hot dogs. Exercise 2.60. The JMP data table **HotDogs.jmp** that is stored in the Sample Data folder contains the data.

10. Poverty and MD's. Exercise 2.64.

11. Life span and body weight. Exercise 2.105.

12. CSDATA. Exercise 2.132.

Chapter 3

Producing Data

3.1 First Steps

There is no need for software in this section.

3.2 Design of Experiments

In a completely randomized design, all the experimental units are allocated at random among the treatments. To accomplish this using JMP INTRO, you create a JMP data table containing all the experimental units (one to a row) and use the **Col Shuffle** function in the **Formula Editor** to randomly select the experimental units for assignment to the treatments.

IPS Example 3.8 Conserving energy

Many utility companies have introduced programs to encourage energy conservation among their customers. An electric company decides to design an experiment to compare three approaches to conservation of energy usage. The company finds 60 single-family residences in the same city willing to participate. You are asked to randomly assign 20 residences to each of the three treatments. A list of the 60 residences (experimental units) is contained in the JMP data table **eg03_008.jmp** located in the Sample Data folder.

1. Open the JMP data table **eg03_008.jmp**.

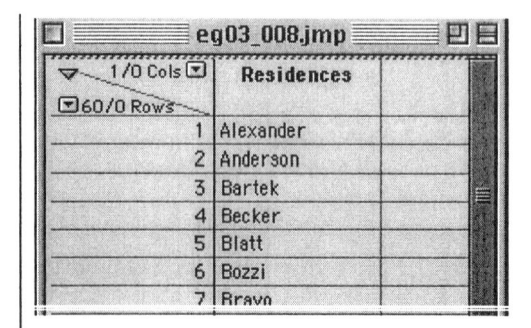

Our approach is first to randomly select the residences one by one.

2. Select **Cols** ⇨ **New Column** and name the column **Order Selected**.

3. Select **New Property** ⇨ **Formula**.

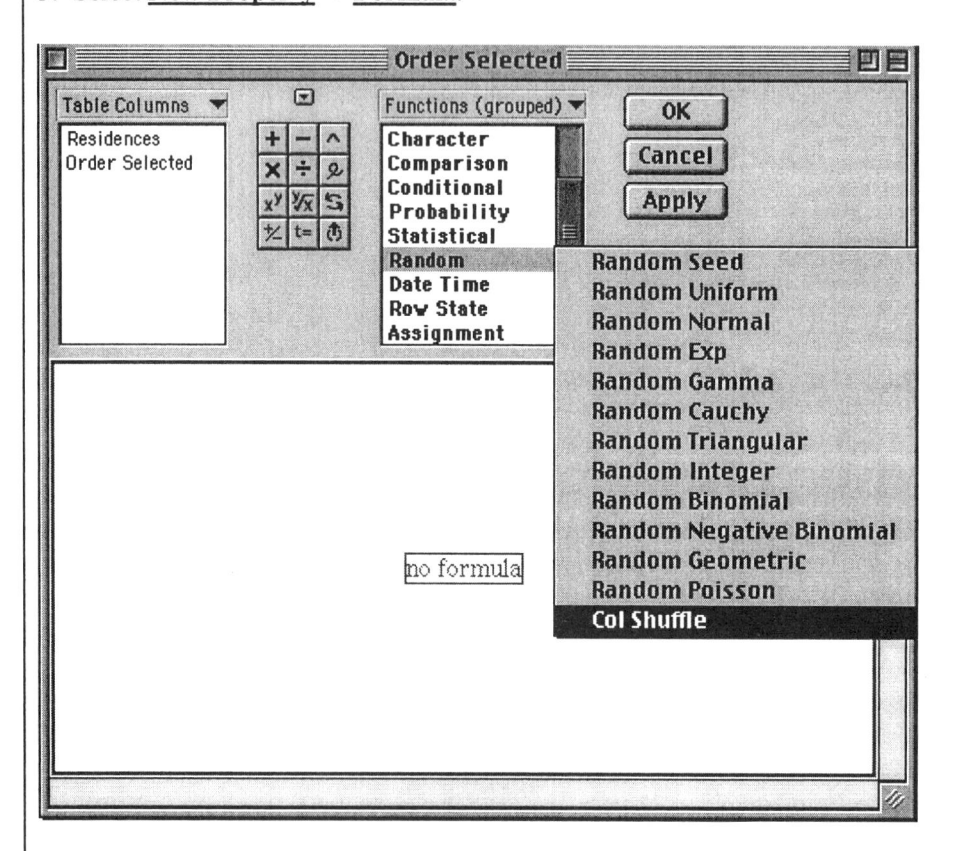

4. Scroll down the list of **Functions (grouped)** and select **Random** ⇨ **Col Shuffle**.

5. Select **OK** and **OK**.

Producing Data

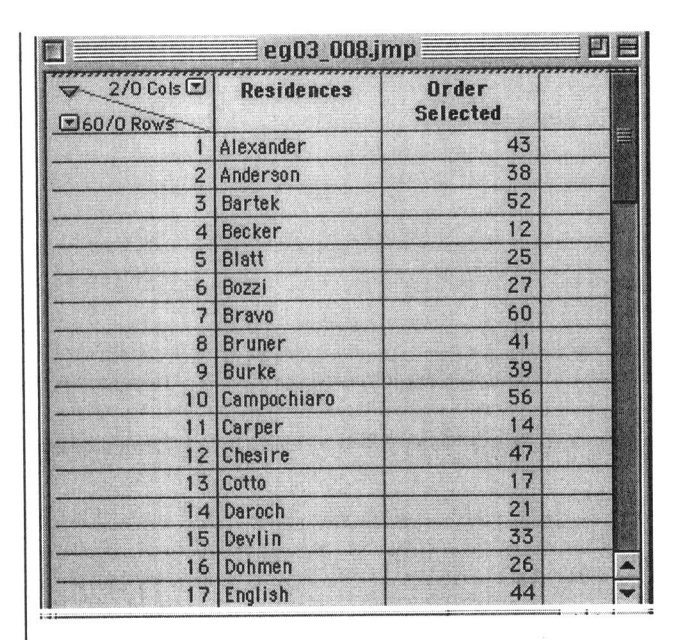

	Residences	Order Selected
1	Alexander	43
2	Anderson	38
3	Bartek	52
4	Becker	12
5	Blatt	25
6	Bozzi	27
7	Bravo	60
8	Bruner	41
9	Burke	39
10	Campochiaro	56
11	Carper	14
12	Chesire	47
13	Cotto	17
14	Daroch	21
15	Devlin	33
16	Dohmen	26
17	English	44

We now create another new variable named **Group**. For the first 20 residences selected, assign it the value 1; for the second 20 residences selected, assign it the value 2; and, for the rest of the residences, assign it the value 3. This is easier to do if you first order the experimental units according to the variable **Order Selected**.

6. Select **Tables** ⇨ **Sort**.

7. Select **Order Selected** and press **By**.

8. Check **Replace Table** and press **Sort** in the ensuing dialog.

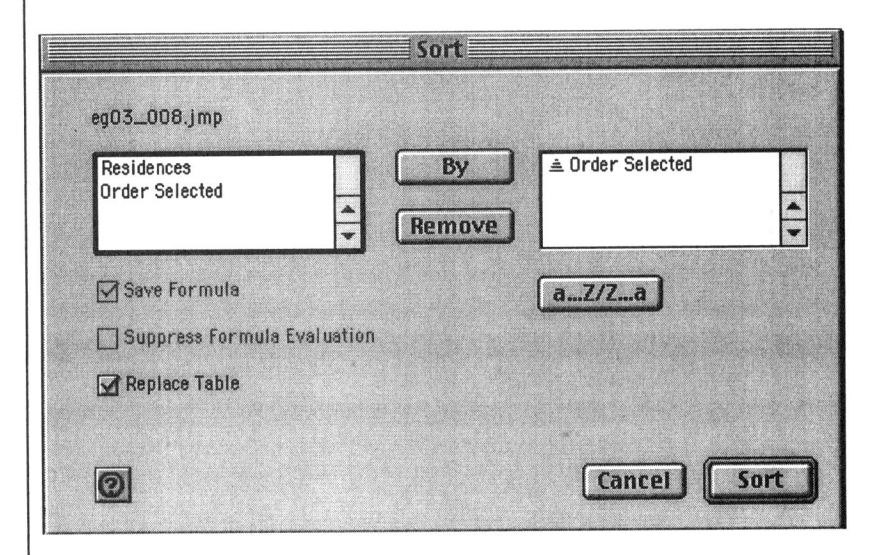

9. Select **Cols** ⇨ **New Column**, name the column **Group**, and press **OK**.

10. Enter a value of **1** in the column **Group** for the first 20 rows, **2** for the next 20 rows, and **3** for the remaining 20 rows.

88   Producing Data

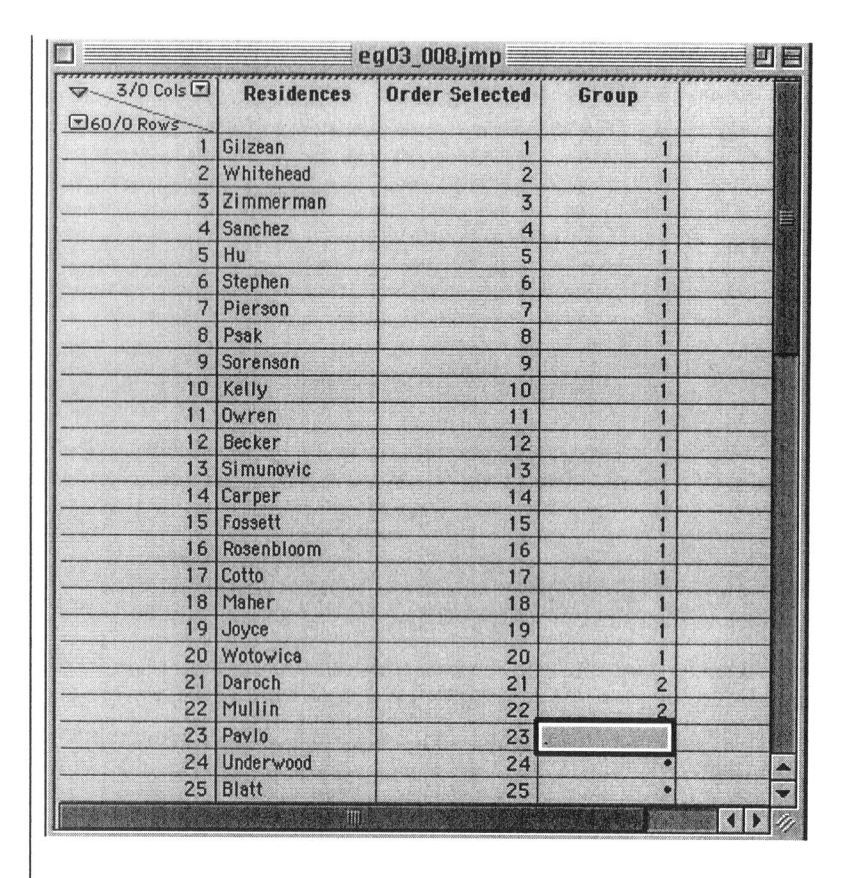

	Residences	Order Selected	Group
1	Gilzean	1	1
2	Whitehead	2	1
3	Zimmerman	3	1
4	Sanchez	4	1
5	Hu	5	1
6	Stephen	6	1
7	Pierson	7	1
8	Psak	8	1
9	Sorenson	9	1
10	Kelly	10	1
11	Owren	11	1
12	Becker	12	1
13	Simunovic	13	1
14	Carper	14	1
15	Fossett	15	1
16	Rosenbloom	16	1
17	Cotto	17	1
18	Maher	18	1
19	Joyce	19	1
20	Wotowica	20	1
21	Daroch	21	2
22	Mullin	22	2
23	Pavlo	23	.
24	Underwood	24	.
25	Blatt	25	.

All that remains is to create a column called **Annual Energy Usage** for the response values, print the table, and give it to the electric company.

Remarks

- The assignment of values to the variable **Group** can be automated with a *formula* using the "If" function and comparison functions.
- JMP IN and the professional versions of JMP have a platform (DOE) for designing experiments, including completely randomized designs, randomized block designs, screening designs, response surface designs, and many other complex and useful designs. Random allocation of the experimental units is automated for each.

3.3 Sampling Design

A simple random sample of n individuals from a population is a subset of n individuals chosen in such a way that all subsets of n individuals have the same chance of being selected. To obtain a simple random sample using JMP, create a JMP data table containing all the individuals in the population and use the **Subset** command on the **Tables** menu.

IPS Example 3.16   How to choose an SRS

An academic department wishes to choose a three-member advisory committee at random from the members of the department. A JMP data table **eg03_016.jmp** that is located in the Sample Data folder contains the names of the 28 faculty. To choose an SRS of size 3 from the 28 faculty,

1. Open the JMP data table **eg03_016.jmp** and notice that it has one variable and 28 rows.

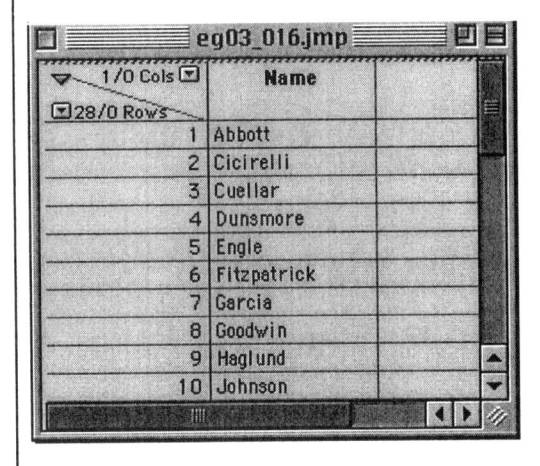

2. Select **Tables** ⇨ **Subset** from the menu bar.

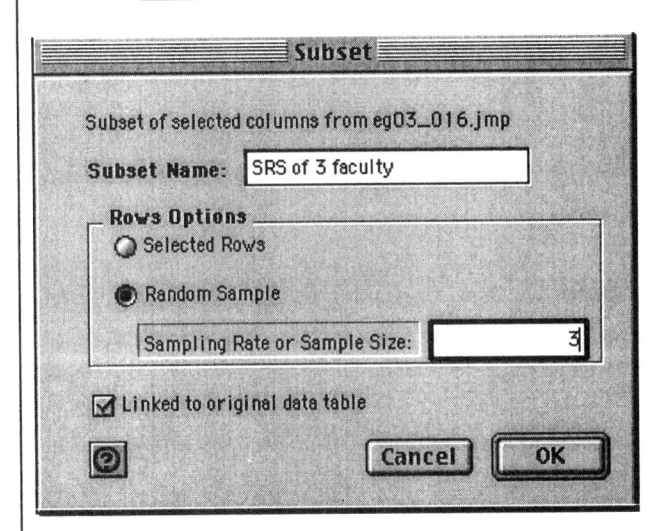

3. Enter a name for the sample, `SRS of 3 Faculty`.

4. Press the button next to `Random Sample` and enter a sample size of **3**.

5. Press **OK**.

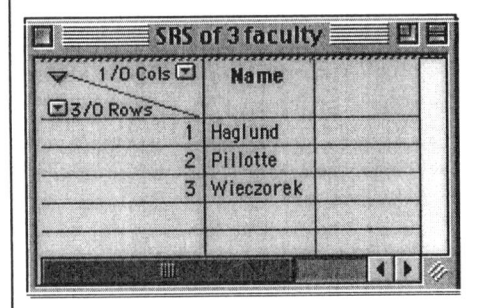

Save the table and send a copy of the resulting simple random sample to the department chair.

3.4 Toward Statistical Inference

To illustrate the sampling variability and sampling distribution of a statistic, the textbook imitates the process of taking many samples. It uses random digits from a table. We will do this in JMP using the built-in random number generator. Let's look at the example in the textbook concerning a market research poll on shopping.

IPS Examples 3.21 and 3.22 Simulation of a sampling distribution

Are attitudes toward shopping changing? A survey by a major market research firm asked a nationwide SRS of adults if they agreed or disagreed that "I like buying new clothes, but shopping is often frustrating and time-consuming." The proportion of the respondents that agrees $p\hat{}$ is calculated. This statistic varies from sample to sample. To illustrate the variability in the sample proportion, let's simulate drawing SRSs of size 100 from the population of all adult U.S. residents. Suppose that, in fact, 60% of the population agree with the statement.

Sampling Variability

The textbook imitates the population by a table of random digits with each entry standing for a person. We will imitate the population with the JMP function *random integer(10)*. Six of the ten integers (1 to 6) stand for people who agree. The remaining four integers, 7 to 10, stand for those who do not. Because all 10 integers are equally likely, the proportion of the resulting population that agrees will be $p = .60$.

Each row of the JMP data table **eg03_021.jmp**, which is located in the Sample Data folder, imitates selecting an individual from the population.

1. Open the JMP data table **eg03_021.jmp**.

Sample Number	Random Integer	Agree
1	4	1
1	7	0
1	8	0
1	10	0
1	9	0
1	10	0
1	9	0
1	1	1
1	9	0
1	9	0
1	1	1
1	1	1
1	10	0
1	5	1
1	2	1
1	5	1
1	4	1
1	6	1
1	1	1
1	10	0
1	2	1

The variable **Agree** holds the individual's response—1 = Yes and 0 = No. The value is based on the value of the *random integer(10)* function given in the second column. Scroll down the table. The first 100 rows represent an SRS of 100 individuals from the population. They contain 64 Yes (1) responses, so $\hat{p} = 0.64$ for this sample.

The table holds 1000 samples of size 100. The variable **Sample Number** indexes the samples. Notice that its value is 1 for the first 100 rows. The second 100 rows represent another SRS of 100 and the variable Sample Number has the value 2 for each of these. The second sample contains exactly 59 individuals who agree with the statement, so $\hat{p} = 0.59$ for this sample.

Sampling Distribution

Let's look at the distribution of the values of $\hat{p}$ from the 1000 SRS of size 100 drawn from this population with $p = 0.60$.

First, we calculate the 1000 sample proportions. The sample proportion for the first sample is the number of ones in the first 100 rows of the column **Agree** divided by 100. The mean of the values in the first 100 rows of the column **Agree** is equivalent to this since the sum of zeros and ones is the number of ones. We will have JMP calculate the sample proportions by calculating the mean of **Agree** for each of the 1000 different samples.

2. Select **Tables** ⇨ **Summary**.

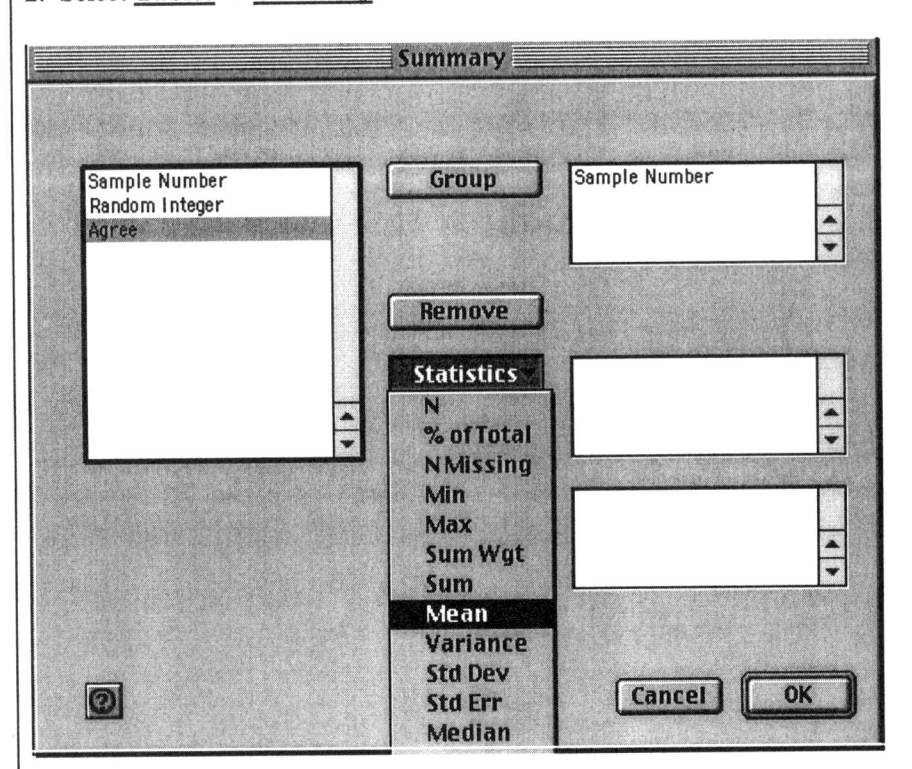

3. Select **Sample Number** and press **Group**.

4. Select **Agree**, and **Statistics** ⇨ **Mean**.

5. Press **OK**.

	Sample Number	N Rows	Sample Proportion
1	1	100	0.64
2	2	100	0.59
3	3	100	0.6
4	4	100	0.73
5	5	100	0.61
6	6	100	0.56
7	7	100	0.52
8	8	100	0.64
9	9	100	0.59
10	10	100	0.64
11	11	100	0.54
12	12	100	0.56
13	13	100	0.64

Rename the variable **Mean(Agree)** to **Sample Proportion** in the resulting data table.

6. Double-click on the variable name **Mean(Agree)** and type `Sample Proportion`.

These 1000 values approximate the sampling distribution of the sample proportion. To explore its shape, center, and spread, use the **Distribution** analysis platform.

7. Select **Analyze** ⇨ **Distribution**.

8. Select the column **Sample Proportion** and press **Y, Columns** and **OK**.

The resulting histogram looks normal and is centered around 0.60. About 95% of the samples have proportions between 0.50 and 0.70.

```
eg03_021.jmp By (Sampl...
Distributions
  Sample Proportion
```

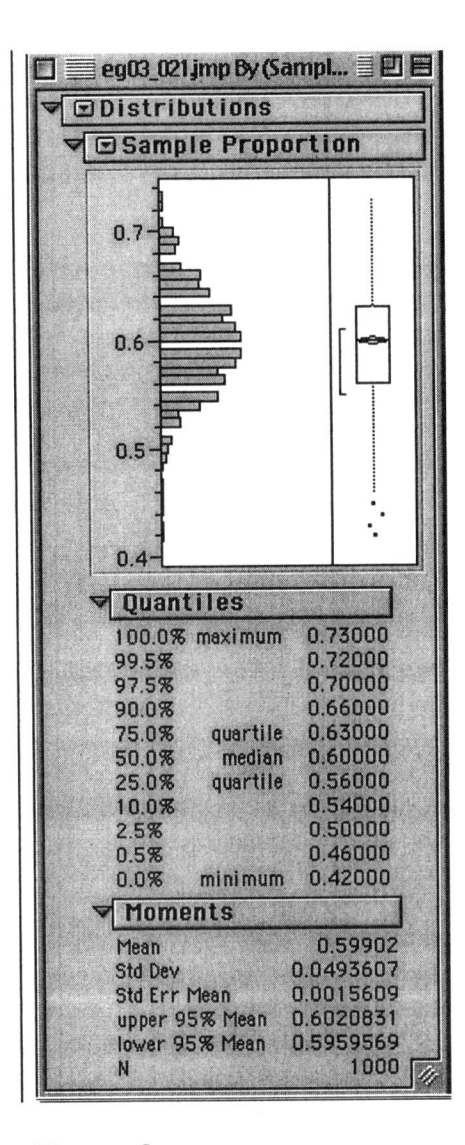

```
Quantiles
100.0%  maximum   0.73000
 99.5%             0.72000
 97.5%             0.70000
 90.0%             0.66000
 75.0%  quartile   0.63000
 50.0%  median    0.60000
 25.0%  quartile   0.56000
 10.0%             0.54000
  2.5%             0.50000
  0.5%             0.46000
  0.0%  minimum   0.42000

Moments
Mean              0.59902
Std Dev           0.0493607
Std Err Mean      0.0015609
upper 95% Mean    0.6020831
lower 95% Mean    0.5959569
N                 1000
```

Remark

JMP INTRO can read JMP data tables of any size. However, it is restricted to saving JMP data tables with 1000 or fewer rows. Therefore, it is difficult to get a good approximation to the sampling distribution of $p\hat{\,}$ for large samples. The professional version of JMP does not place any restrictions on the size of a data table; it can be used to simulate sampling distributions for large sample sizes.

3.5 Summary

Activity	Command
Simple random samples	**Tables** ⇨ **Subset**
Completely randomized designs	**Cols** ⇨ **New Column** ... **Formula** ...**Col Shuffle**

3.6 Exercises

Use JMP to carry out the randomization for the following exercises from the textbook. For each exercise, you will need to create a new JMP data table containing either the members of the population or a list of available experimental units. See Section 0.2.2 in Chapter 0 for details on how to create a new JMP data table.

1. Relief for chronic tension-type headaches. Exercise 3.13.

2. Maintaining normal body temperature during surgery. Exercise 3.14.

3. Sealing food packages. Exercise 3.15.

4. Comparing methods of teaching seventh-graders math. Exercise 3.27.

5. Comparing weight reduction regimens. Exercise 3.32.

6. Apartment living. Exercise 3.40.

7. Attitudes of minority managers. Exercise 3.41.

8. Stratified random sampling. Exercise 3.4.7. Hint: Select an SRS of 4 students. Then select an SRS of 2 faculty members. These six people are a stratified random sample.

9. Sampling a field for egg masses of a harmful insect. Exercise 3.69. Open the JMP data table **ex 03_069.jmp** that is located in the Sample Data folder. Examine the formula for the variables **Random Integer** and **Present**. The JMP *random integer(5)* function is used to imitate the population of all yard-square areas. One of the five integers (1) stands for a yard-square area in which egg masses are present. The remaining four integers (2 to 5) stand for those in which egg masses are absent. Simulate the results of taking 100 samples of size 10 (1000 samples if you are using the professional version of JMP). Answer the questions in part (b) of the exercise. Make a normal quantile plot of the 100 sample proportions. Is the sampling distribution approximately normal?

10. Are attitudes toward shopping changing? If you have access to the professional version of JMP, you can simulate the sampling distribution of $p\hat{}$ for SRSs of size 2500. The JMP data table **eg03_022.jmp** that is located in the Sample Data folder will generate SRSs of size 2500. Use the **Add Rows** command on the **Rows** menu to generate 1000 samples of size 2500.

11. Sampling populations with $p = 0.1$, $p = 0.3$, and $p = 0.5$. Exercise 3.92 (This is best done using the professional version of JMP because of the size of the sample.) Modify the formulae in the data table **eg03_021.jmp** to accommodate the different values of the parameter p. Use the *random integer(10)* function.

Chapter 4

Probability—The Study of Randomness

This chapter concerns information about probability that is useful in understanding and managing randomness. The JMP probability functions, **Binomial Distribution** and **Binomial Probability**, compute binomial probabilities; these can be accessed from the *Formula Editor* in a manner analogous to the example for normal probabilities in Section 1.3 in Chapter 1. However, the calculations in this chapter are best performed using a calculator.

Chapter 5

Sampling Distributions

5.1 Sampling Distributions for Proportions

The location, spread, and shape of the distribution of the sample proportion can be obtained by the laws of probability. If $\hat{p}$ represents the sample proportion of successes in an SRS of size n drawn from a large population having population proportion p of successes, then the mean and standard deviation of $\hat{p}$ are

$$\mu_{\hat{p}} = p$$

$$\sigma_{\hat{p}} = \sqrt{\frac{p(1-p)}{n}}$$

Example 5.1 Are attitudes toward shopping changing?

An SRS of 100 U.S. adults asks whether they agreed or disagreed that "I like buying new clothes, but shopping is often frustrating and time-consuming." The proportion of the sample that agreed is to be calculated. It is a statistic and varies from sample to sample. The mean and standard deviation of this sample proportion are

$$\mu_{\hat{p}} = p = 0.6$$

$$\sigma_{\hat{p}} = \sqrt{\frac{p(1-p)}{n}} = \sqrt{\frac{(0.6)(0.4)}{100}} = 0.048989$$

In Chapter 3, we conducted a simulation to imitate the process of taking 1000 SRSs of size 100 from a population in which 60% of the individuals agreed with the statement. The resulting distribution of the 1000 sample proportions approximates the sampling distribution of $\hat{p}$.

We can compare the mean and standard deviation of the simulated distribution with the values that we have just calculated. In Chapter 3, the **Distribution** command gave

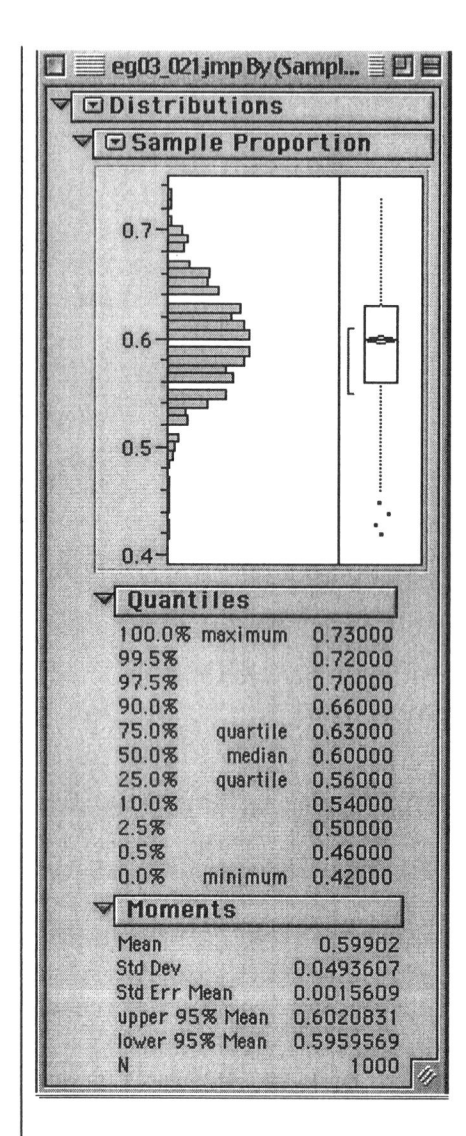

```
┌─────────────────────────────────────┐
│ □ ≡ eg03_021.jmp By (Sampl...  ≡ ▣ ▣│
│ ▽ □ Distributions                   │
│   ▽ □ Sample Proportion             │
│   ┌─────────────────────────┐       │
│   │                         │       │
│   │  0.7-                   │       │
│   │                         │       │
│   │                         │       │
│   │  0.6-          [ ▣ ]    │       │
│   │                         │       │
│   │  0.5-                   │       │
│   │                     ⋮   │       │
│   │  0.4-                   │       │
│   └─────────────────────────┘       │
│ ▽ Quantiles                         │
│    100.0%  maximum   0.73000        │
│     99.5%            0.72000        │
│     97.5%            0.70000        │
│     90.0%            0.66000        │
│     75.0%  quartile  0.63000        │
│     50.0%  median    0.60000        │
│     25.0%  quartile  0.56000        │
│     10.0%            0.54000        │
│      2.5%            0.50000        │
│      0.5%            0.46000        │
│      0.0%  minimum   0.42000        │
│ ▽ Moments                           │
│    Mean              0.59902        │
│    Std Dev           0.0493607      │
│    Std Err Mean      0.0015609      │
│    upper 95% Mean    0.6020831      │
│    lower 95% Mean    0.5959569      │
│    N                 1000           │
└─────────────────────────────────────┘
```

The empirical results (mean = 0.59902 and standard deviation = 0.0493607) are very close to the true values. In addition, the shape of the histogram is consistent with the DeMoivre-Laplace theorem—when n is large the sampling distribution of the sample proportion is approximately normal. As an exercise, you will be asked to construct a normal quantile plot to confirm this.

5.2 The Sampling Distribution of a Sample Mean

The laws of probability can be used to derive the mean and standard deviation of the sampling distribution of $\bar{x}$ of an SRS of size n from a population with mean μ and standard deviation σ. The mean and standard deviation of $\bar{x}$ are

$$\mu_{\bar{x}} = \mu$$

$$\sigma_{\bar{x}} = \frac{\sigma}{\sqrt{n}}$$

The *Central Limit Theorem* tells us that the distribution of $\bar{x}$ will be approximately normal no matter what the shape of the population distribution as long as the sample size is large enough (and the standard deviation is finite).

To better understand these results and the concept of a sampling distribution, we will simulate the sampling distributions for means of SRSs of different sizes from a very skewed distribution.

Example 5.2 Simulating the sampling distribution of a sample mean

1. Open the JMP data table **CLT.jmp** that is located in the Sample Data folder.

2. Select **Rows** ⇨ **Add Rows**.

3. Type **1000** and press **OK**.

Here is a subset of one possible data table. Since the data values are generated at random, your table will have different values.

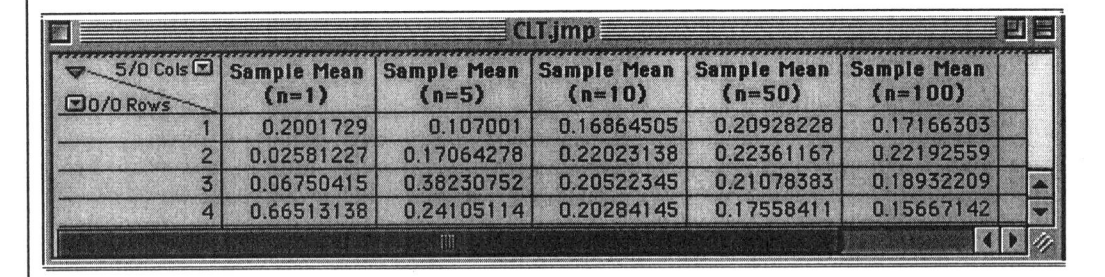

	Sample Mean (n=1)	Sample Mean (n=5)	Sample Mean (n=10)	Sample Mean (n=50)	Sample Mean (n=100)
1	0.2001729	0.107001	0.16864505	0.20928228	0.17166303
2	0.02581227	0.17064278	0.22023138	0.22361167	0.22192559
3	0.06750415	0.38230752	0.20522345	0.21078383	0.18932209
4	0.66513138	0.24105114	0.20284145	0.17558411	0.15667142

Each row imitates selecting an SRS from a very skewed distribution. There are five columns. The first column contains the mean for each sample of size 1. Hence, these are random values from the population distribution and their distribution will approximate the population distribution. Let's use JMP to examine the distribution.

4. Select **Analyze** ⇨ **Distribution.**

5. Select **Sample Mean (n=1)** and press **Y, Columns** and **OK**.

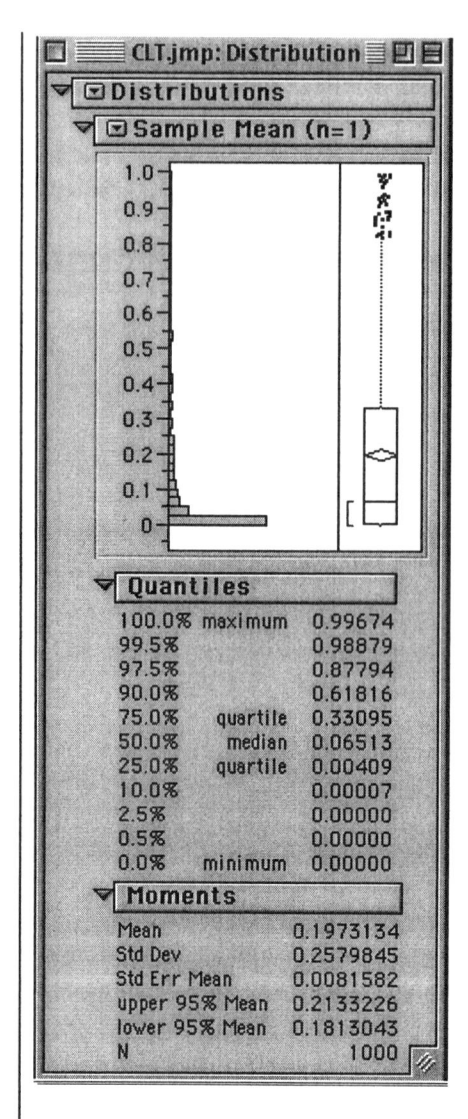

```
CLT.jmp: Distribution
Distributions
  Sample Mean (n=1)

        1.0
        0.9
        0.8
        0.7
        0.6
        0.5
        0.4
        0.3
        0.2
        0.1
          0

  Quantiles
    100.0%  maximum   0.99674
     99.5%            0.98879
     97.5%            0.87794
     90.0%            0.61816
     75.0%   quartile 0.33095
     50.0%   median   0.06513
     25.0%   quartile 0.00409
     10.0%            0.00007
      2.5%            0.00000
      0.5%            0.00000
      0.0%  minimum   0.00000
  Moments
    Mean              0.1973134
    Std Dev           0.2579845
    Std Err Mean      0.0081582
    upper 95% Mean    0.2133226
    lower 95% Mean    0.1813043
    N                      1000
```

Thus, the mean μ and the standard deviation σ of the population distribution are approximately 0.1973134 and 0.2579845, respectively. (Of course, since the data values are generated at random, your results will differ slightly). You can also see that the distribution is strongly skewed toward larger numbers. Waiting times for service (e.g., a highway tollbooth lane) often follow such a distribution.

Now, let's look at the distributions of the sample mean for SRSs of sizes 5, 10, 50, and 100.

6. Select **Analyze** ⇨ **Distribution**.

7. Select **Sample Mean (n=5), ..., Sample Mean (n=100)** and press **Y, Columns** and **OK**.

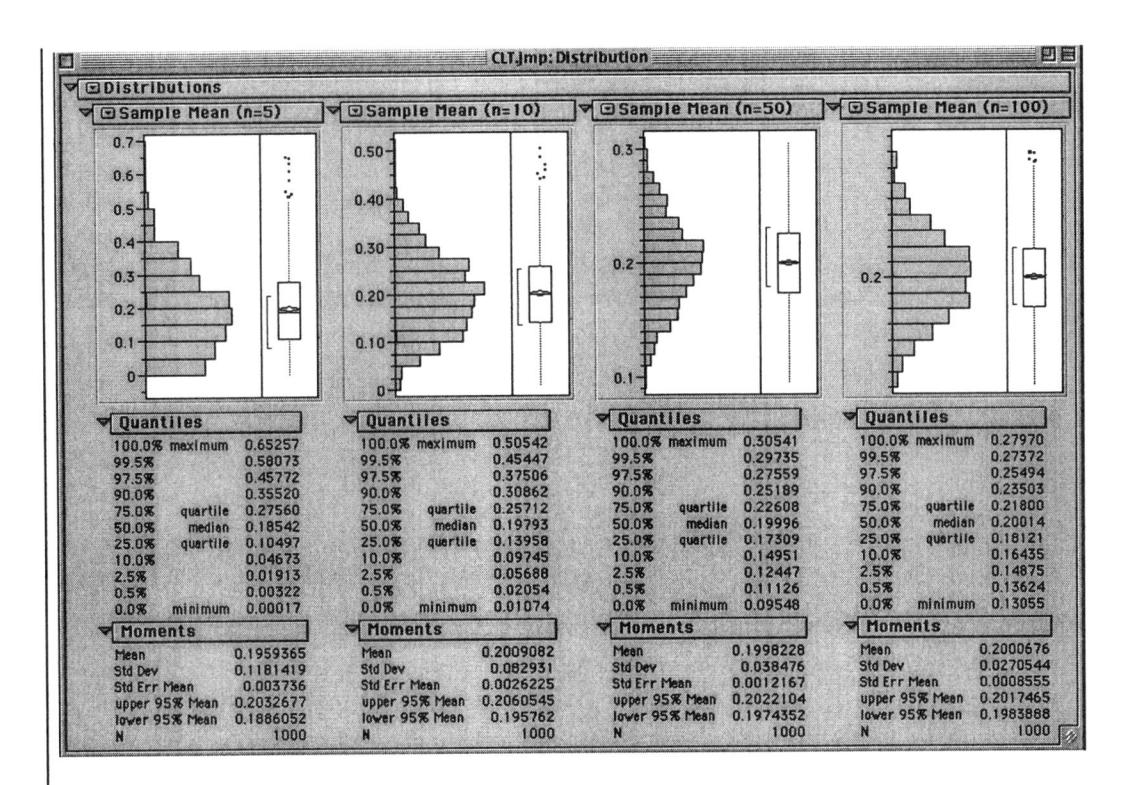

[Screenshot: CLT.jmp Distribution window showing four Sample Mean distributions for n=5, n=10, n=50, n=100, each with histogram, box plot, Quantiles, and Moments panels.]

	Sample Mean (n=5)	Sample Mean (n=10)	Sample Mean (n=50)	Sample Mean (n=100)
Quantiles				
100.0% maximum	0.65257	0.50542	0.30541	0.27970
99.5%	0.58073	0.45447	0.29735	0.27372
97.5%	0.45772	0.37506	0.27559	0.25494
90.0%	0.35520	0.30862	0.25189	0.23503
75.0% quartile	0.27560	0.25712	0.22608	0.21800
50.0% median	0.18542	0.19793	0.19996	0.20014
25.0% quartile	0.10497	0.13958	0.17309	0.18121
10.0%	0.04673	0.09745	0.14951	0.16435
2.5%	0.01913	0.05688	0.12447	0.14875
0.5%	0.00322	0.02054	0.11126	0.13624
0.0% minimum	0.00017	0.01074	0.09548	0.13055
Moments				
Mean	0.1959365	0.2009082	0.1998228	0.2000676
Std Dev	0.1181419	0.082931	0.038476	0.0270544
Std Err Mean	0.003736	0.0026225	0.0012167	0.0008555
upper 95% Mean	0.2032677	0.2060545	0.2022104	0.2017465
lower 95% Mean	0.1886052	0.195762	0.1974352	0.1983888
N	1000	1000	1000	1000

First, look at the means for each distribution and compare them to the mean of the first column, which approximates the population mean. Notice that they are almost equal.

Second, consider the standard deviations of these distributions, 0.1181419, 0.082931, 0.038476, and 0.0270544, respectively. Note that they become progressively smaller as expected. According to the formula, these standard deviations should each be approximately $1/\sqrt{n}$ times the population standard deviation. For the simulation shown here, the population standard deviation is approximately 0.2579845 and so the standard deviations of these sampling distributions should have been approximately 0.115374, 0.081582, 0.036485, and 0.025798, respectively. They are remarkably close to the corresponding empirical standard deviations. How do the empirical standard deviations of your simulation compare with the theoretical results?

Now look at the shapes of the distributions. Notice that they tend to look more like a normal curve as the sample size increases even though the population distribution is very skewed. That's what the famous *Central Limit Theorem* says!

5.3 Exercise

Are attitudes toward shopping changing? In Examples 3.21 and 5.1, you simulated the sampling distribution of the sample proportion of the survey respondents who agreed with the statement. Use JMP to construct a normal quantile plot to confirm that the normal approximation is reasonable.

Chapter 6

Introduction to Inference

6.1 Estimating with Confidence

Calculations in this section are best done with a calculator. The calculations are not difficult and reinforce their dependence on the sampling distribution of the sample mean. JMP will be used to compute confidence interval estimates for the mean when the population standard deviation is unknown; this is the usual case (see Chapter 7).

You have learned that a 95% confidence interval for the mean is arrived at "by a method that gives correct results 95% of the time." You have also learned that

- an interval calculated from an SRS of size 100 will be narrower than one based on an SRS of size 20,
- an interval for a population with a standard deviation of 2 will be wider than one with a standard deviation of 1, and
- a 99% confidence interval is wider than a 95% confidence interval.

To better understand these notions, we will illustrate them using a JMP script that is in the Sample Scripts folder with the JMP INTRO application.

1. Select **File** ⇨ **Open**.

2. Select the folder **Sample Scripts** on the Open: JMP INTRO dialog.

3. Select **All Readable Documents** from the menu below the list of files.

4. Select the script **Confidence.jsl** and press **Open**.

A text window containing the script will open. To run the script,

5. Select **Edit** ⇨ **Run Script.**

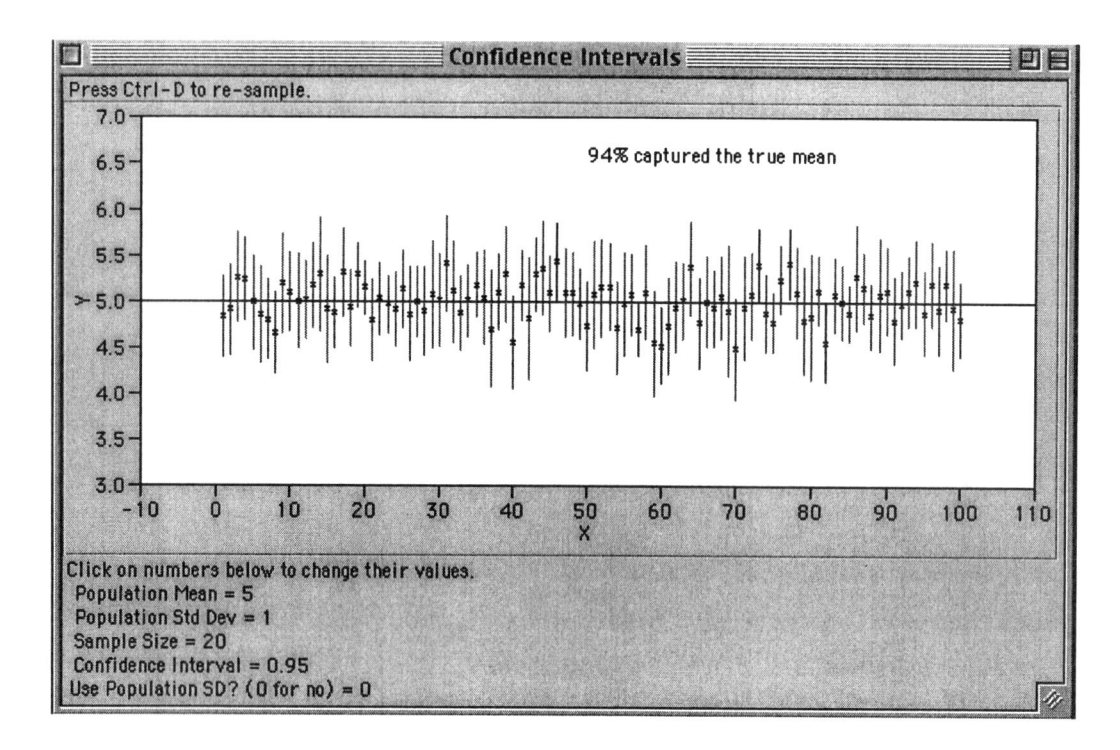

[Screenshot of "Confidence Intervals" window showing a plot with "94% captured the true mean" annotation. The plot shows 100 confidence intervals plotted vertically around a horizontal line at y=5.0, with y-axis ranging from 3.0 to 7.0 and x-axis from -10 to 110. Below the plot:
"Click on numbers below to change their values.
Population Mean = 5
Population Std Dev = 1
Sample Size = 20
Confidence Interval = 0.95
Use Population SD? (0 for no) = 0"]

The **Confidence Intervals** window contains a plot that simulates taking 100 SRSs from a population with known mean $\mu = 5$ and calculating the 100 associated 95% confidence intervals for these samples. The horizontal axis indexes the samples; the values of the variable are on the vertical axis. Red intervals indicate that the interval did not include the population mean. (These correspond to the intervals that are entirely above or below the horizontal line at 5.0.) The percent that capture the true mean is displayed in the graph. We expect that percent to be near 95% and in the long run to converge to 95%. For the simulation shown here, six of the samples have intervals that do not include the population mean. Hence, 94% capture the true mean.

Let's see the effect of increasing the sample size to 100 from 20. First, specify that the population standard deviation should be used.

6. Click on the last **0** in the last line, type **1**, and press the **return key**.

7. Click on the sample size **20**, type **100**, and press the **return key**.

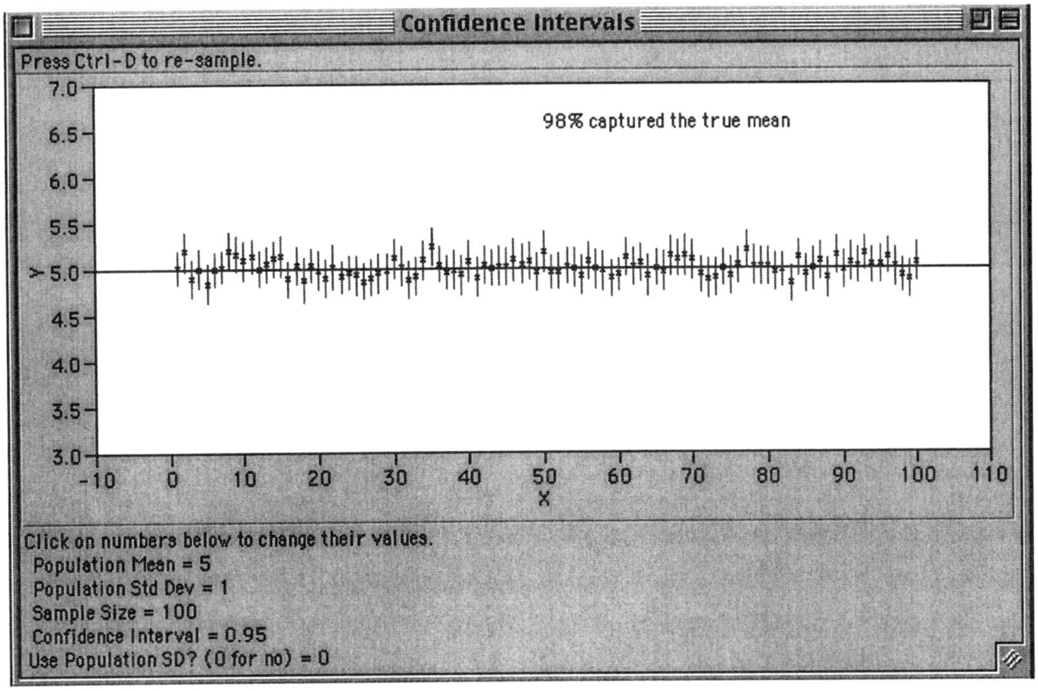
Confidence Intervals
Press Ctrl-D to re-sample.
98% captured the true mean
7.0
6.5
6.0
5.5
>5.0
4.5
4.0
3.5
3.0
-10 0 10 20 30 40 50 60 70 80 90 100 110
X
Click on numbers below to change their values.
Population Mean = 5
Population Std Dev = 1
Sample Size = 100
Confidence Interval = 0.95
Use Population SD? (0 for no) = 0

The intervals are noticeably narrower, as expected. Now let's see the effect of increasing the standard deviation.

8. Return to a sample size of 20 by clicking on the sample size of **100** and typing **20**.

9. Click on the Population Std Dev **1** and type **2**.

As expected, these intervals are noticeably wider than the ones with a standard deviation of 1.

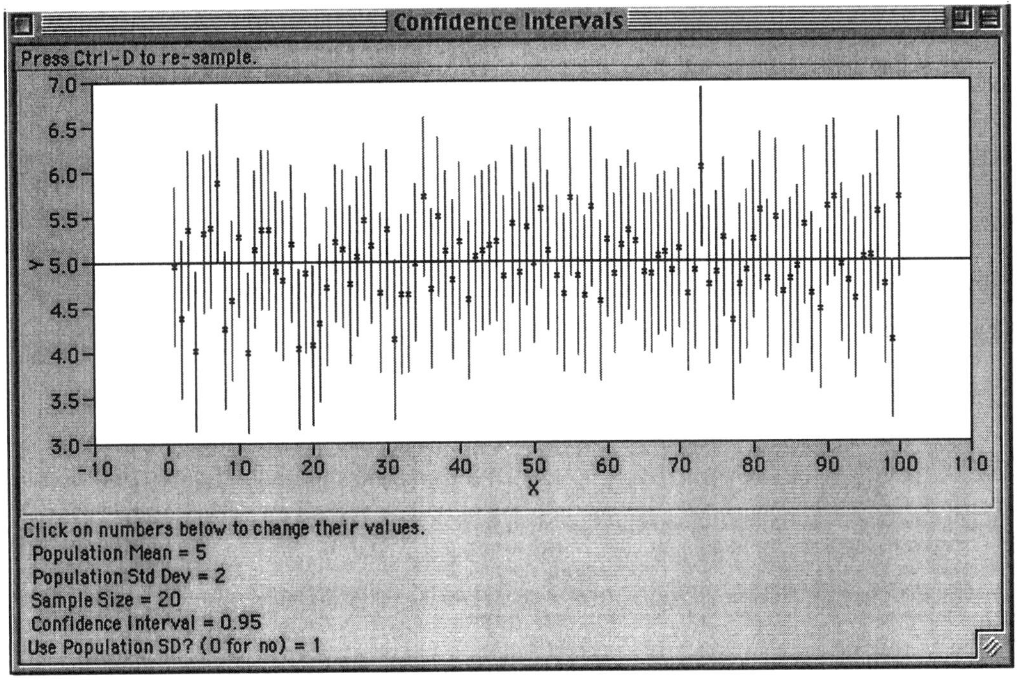
Confidence Intervals
Press Ctrl-D to re-sample.
7.0
6.5
6.0
5.5
>5.0
4.5
4.0
3.5
3.0
-10 0 10 20 30 40 50 60 70 80 90 100 110
X
Click on numbers below to change their values.
Population Mean = 5
Population Std Dev = 2
Sample Size = 20
Confidence Interval = 0.95
Use Population SD? (0 for no) = 1

Now let's see the effect of increasing the confidence level to 99%.

10. Click on the confidence level of **0.95**, type **0.99,** and press the **return key**.

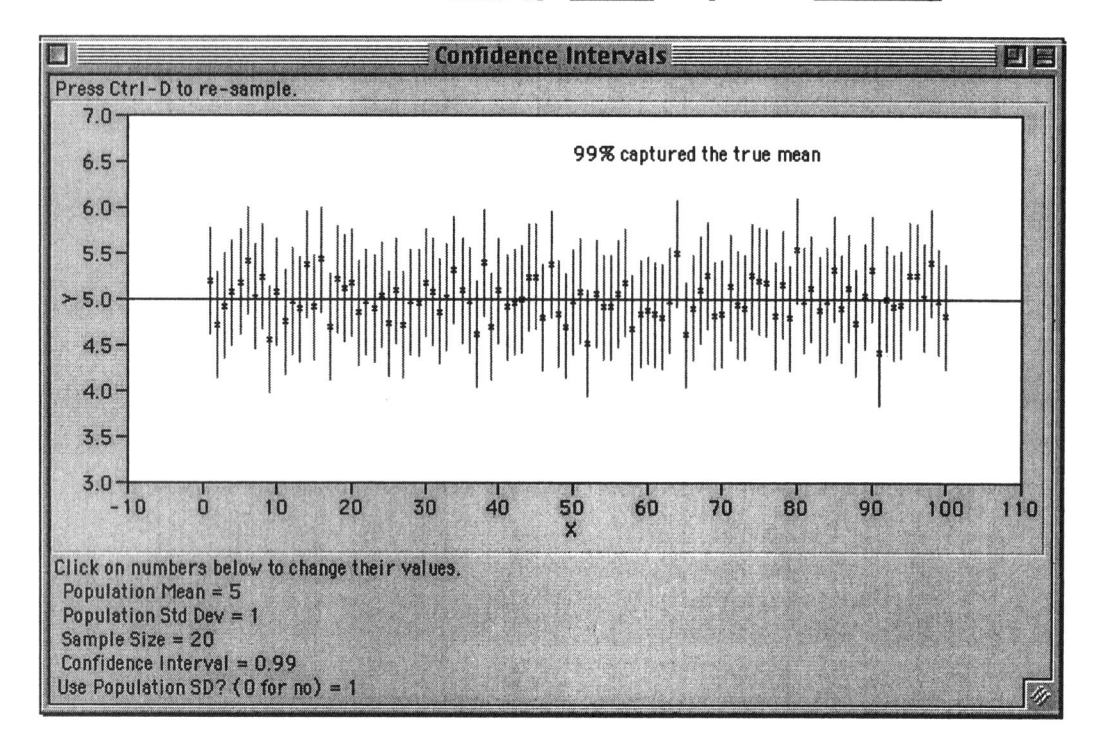

These 99% confidence intervals tend to be wider than the 95% confidence intervals for SRSs of size 20.

Why aren't they called probability intervals? In addition, what does it mean to say, "The confidence level shows how *confident* we are that the *procedure* will catch the true population mean?"

On the one hand, when we calculate a confidence interval, we normally get one and only one interval, say 4.2 to 5.1. In addition, that interval either contains the unknown mean or not. Since the mean is unknown but not random and since our interval [4.2, 5.1] is fixed, there is no random experiment. So the probability of our interval containing the mean is either 1 or 0 depending on whether or not the mean is included in the interval. Hence, we **cannot** refer to them as probability intervals.

On the other hand, the method by which a confidence interval is constructed provides that if you repeatedly take SRSs of the same size and calculate 95% confidence intervals, then 95% of those intervals will contain the population mean. To see this, simulate taking many 95% confidence intervals using the **Confidence Intervals** window.

11. Reset the sample size to **10** and the confidence level to **0.95**.

12. Resample by pressing **ctrl-D** under Windows and **Command-D** on a Macintosh.

Repeat this and calculate the proportion of intervals that capture the population mean after 100 intervals, 200 intervals, 500 intervals, and 1000 intervals. While the percent will vary from one set of 100 intervals to another, it should be close to 95% and the overall percent should approach 95% as more intervals are generated.

That's what we mean when you say, "The confidence level shows how *confident* we are that the *procedure* will catch the true population mean."

6.2 Tests of Significance

Tests of significance are easy to perform in JMP. Since only one variable is involved, we use the **Distribution** platform. The **Test Mean** command performs *z*-tests.

IPS Example 6.6 Tim Kelley's weight

Tim Kelley has a driver's license that gives his weight as 187 pounds. He did not change this information the last time that he renewed his license although he suspects that he may have gained a little weight over the past few years. He records his weight on a regular basis. The four readings for last month were 190.5, 189.0, 195.5, and 187.0. Assume that the standard deviation σ of his weight measurements is 3 pounds. Are these observations compatible with a true mean weight μ of 187 pounds? To answer this question, use a test of significance on the alternative hypothesis that

$$H_a: \mu > 187$$

1. Select **File** ⇨ **New**.

Create a JMP data table with one variable named **Weight** and four rows. Save the data table as **KelleyWt.jmp** for later use. (See Section 0.2.2 in Chapter 0 for details on creating a new JMP data table.)

2. Select **Analyze** ⇨ **Distribution** to look at the distribution of **Weight**.

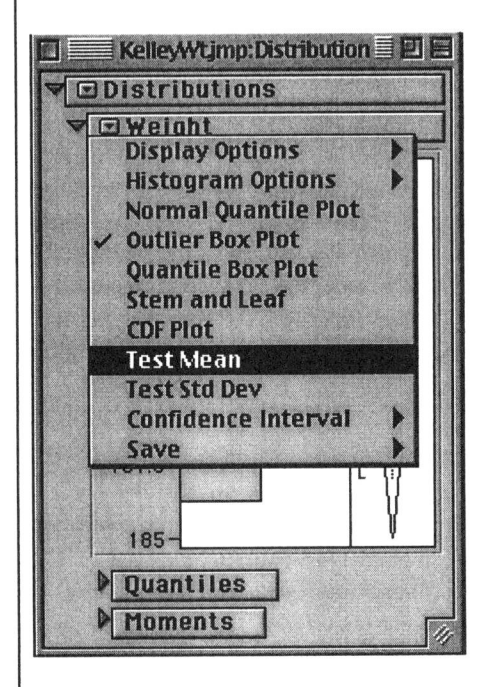

3. Click on the red triangle next to the title bar for the **Weight** report and select **Test Mean** from the menu that opens.

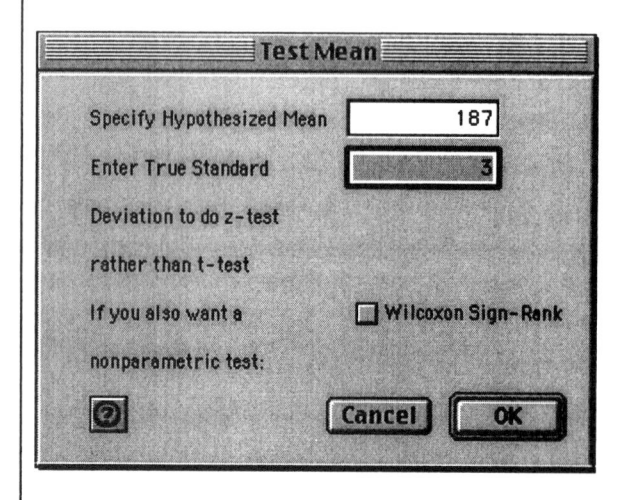

4. Type **187** in the `Specify Hypothesized Mean` field.

5. Type **3** for the `True Standard deviation`.

6. Select **OK**.

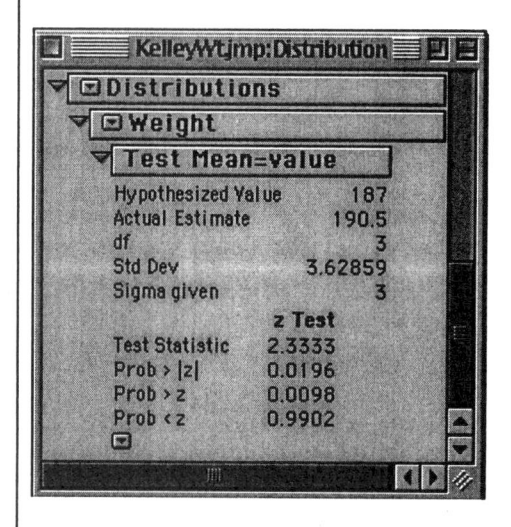

Notice that JMP calculates the value of the *z* **Test** statistic for the sample, *z* = 2.33, and provides three *P*-values, one for each of the three possible alternative hypotheses.

Alternative Hypothesis		*P*-value in JMP
$\mu \neq 187$	Two-tailed	Prob > |z|
$\mu > 187$	Upper tail	Prob > z
$\mu < 187$	Lower tail	Prob < z

Since you wish to test that H_a: $\mu > 187$, use the *P*-value for an upper-tailed test, 0.0098. Thus, there is strong evidence that Tim's mean weight has increased.

P-values

To see how a *P*-value measures the difference between the hypothesized mean and the sample mean, let's look at a graph.

7. Click on the red triangle at the bottom of the **Test Mean=Value** report and select **PValue animation** from the menu that opens.

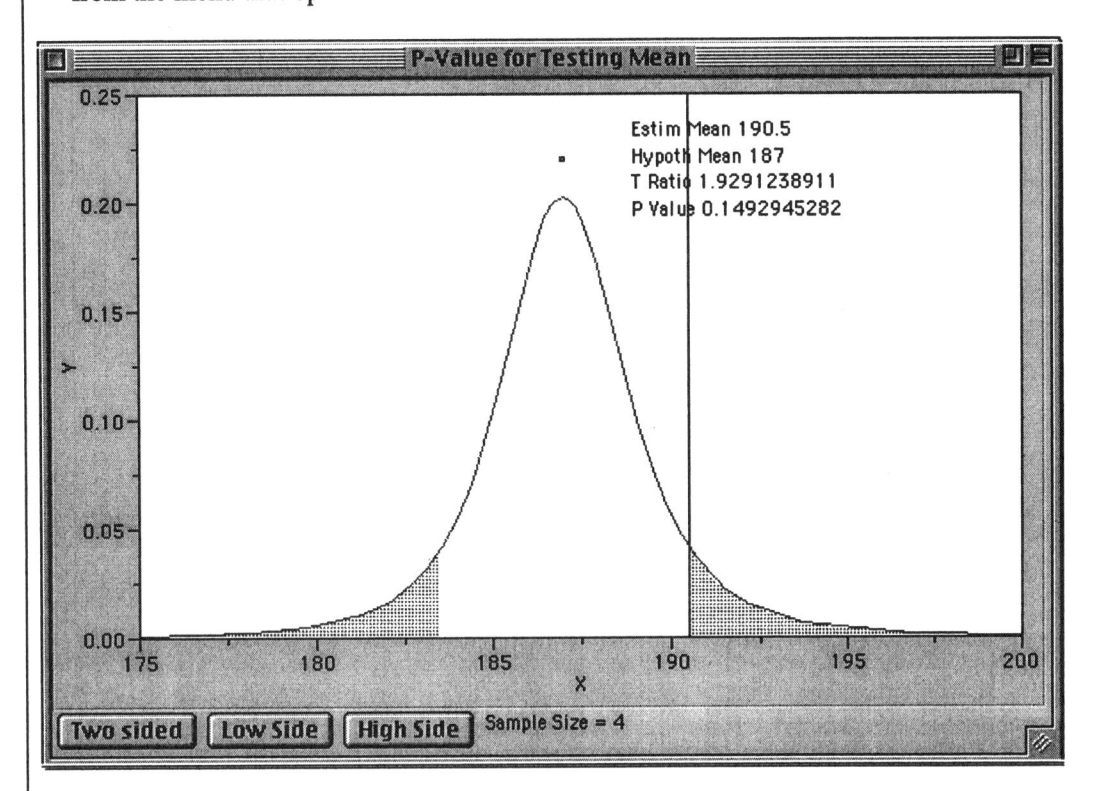

[Figure: P-Value for Testing Mean window showing a bell-shaped curve with both tails shaded. Estim Mean 190.5, Hypoth Mean 187, T Ratio 1.9291238911, P Value 0.1492945282. Buttons: Two sided, Low Side, High Side. Sample Size = 4.]

Examine the graph that opens. Notice that both tails are shaded and that the shaded area is associated with a two-sided alternative hypothesis. Our alternative hypothesis is a one-sided one.

8. Press the **High Side** button. Notice that only the upper tail is shaded.

The *P*-values calculated in the animation window do not assume that the population standard deviation is known. Therefore, they will not correspond exactly to the *P*-values that we calculate in this chapter. Nonetheless, their order of magnitude will be helpful.

Let's see what happens to the *P*-value as we move the hypothesized mean away from the observed (sample) mean.

9. Press the handle (a small square) directly above the peak of the bell-shaped curve and drag it to the left.

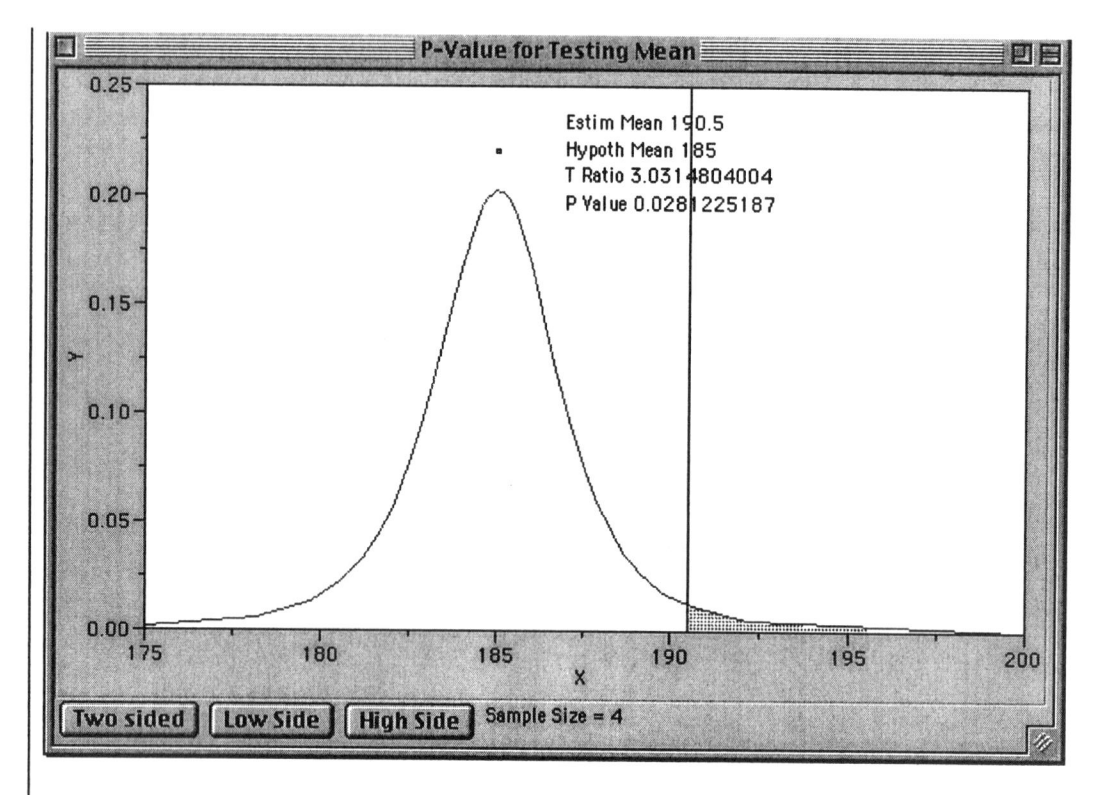

[Screenshot: P-Value for Testing Mean window showing Estim Mean 190.5, Hypoth Mean 185, T Ratio 3.0314804004, P Value 0.0281225187, Sample Size = 4]

The entire curve moves and the hypothesized mean becomes smaller. The *P*-value becomes increasingly smaller as the hypothesized mean moves further away from 187. It changes from about 0.075 to 0.028 as the hypothesized mean goes from 187 to 185.

6.3 Summary

The focus of this chapter is to introduce you to confidence intervals and significance tests and the reasoning behind them. Although JMP software can be useful in some ways, most of the problems can be done by hand or with a calculator.

All graphs and statistical computations in this chapter are performed in the first platform **Distribution** of the **Analyze** menu.

Graph/Computation	Command
Significance tests	**Analyze** ⇨ **Distribution** ⇨ **Test Mean**

6.4 Exercises

1. The JMP script **Confidence.jsl** allows you to quickly simulate 100 confidence intervals at a time. Run the script 30 times for a 95% confidence interval with known population standard deviation equal to 1 and record the number of intervals that capture the true mean. Make a stemplot of the 30 numbers. (Hint: Put them into a JMP data table and use the Distribution command.) Describe the results. If you were to repeat this experiment a very large number of times, what would you expect the mean number of intervals that cover the true mean to be?

Introduction to Inference

2. Repeat the previous exercise for 80% confidence.

Use JMP to help carry out the following exercises from the textbook. Data for the exercises can be imported from text files on the IPS CD-ROM with names corresponding to the exercise number or associated table number and a suffix of *.txt*.

3. Local school district's reading scores. Exercise 6.49.

4. Weights of male runners. Exercise 6.62.

5. In the previous exercise, use the p-value animation to see how the *P*-value changes as the hypothesized mean changes from 61.3 kg to 65 kg to 67 kg.

6. SAT-M scores for California high school seniors. Exercise 6.107.

Chapter 7

Inference for Distributions

This chapter emphasizes the practice of statistical inference. Population standard deviations are no longer unrealistically assumed to be known; we learn about a new family of sampling distributions, the Student t-distributions; and we learn to compare two distributions.

For inference on a single variable, we employ the **Distribution** platform in JMP. We use the **Fit Y by X** platform to compare the means of two populations because this is equivalent to determining whether two variables are related. Most of the skills that you learned in Chapter 1 for examining distributions and in Section 2.1 in Chapter 2 for examining the relationship of a quantitative response variable and a categorical explanatory variable will be used here.

7.1 Inference for the Mean of a Population

Since we are concerned with one variable, the **Distribution** analysis platform is used. When the population standard deviation is unknown, as is usually the case, JMP calculates the sample standard deviation and uses it in place of the true standard deviation in the test statistic. The t-statistic and associated P-value are automatically calculated for tests of significance and the appropriate quantile of the t-distribution is used for confidence intervals. The following examples illustrate this.

7.1.1 The One-Sample t Confidence Interval

IPS Example 7.4 Passage time of light

One of the first sets of data that we examined was Simon Newcomb's measurement of the passage time of light. Let's find a 99% confidence interval for μ, the mean of the distribution of measurements from which Newcomb's data are a sample.

1. Open the JMP data table containing Newcomb's data that you saved in Chapter 1. If you have misplaced the data table, you can import the file **ta01_001.txt**. Name the column of data values **Passage Time**.

You may recall from Section 1.3 in Chapter 1 that there were two unusually small measurements in Newcomb's data. We will examine the data without these outliers and verify that the remaining measurements follow a normal distribution.

2. Click on row **6** and ctrl-click (command-click on the Macintosh) on row **10** to select the two outliers.

3. Select **Rows** ⇨ **Exclude/Unexclude** to change the row state of these measurements and exclude them from analyses. (See Section 0.3.2 in Chapter 0.)

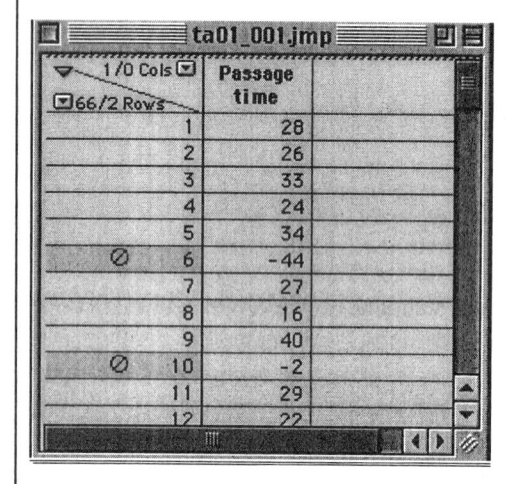

4. Select **Analyze** ⇨ **Distribution**.

5. Select **Passage Time** and press the **Y, Columns** and **OK** buttons just as we did in Chapter 1.

6. Press the red triangle next to **Passage Time** and select **Normal Quantile Plot**.

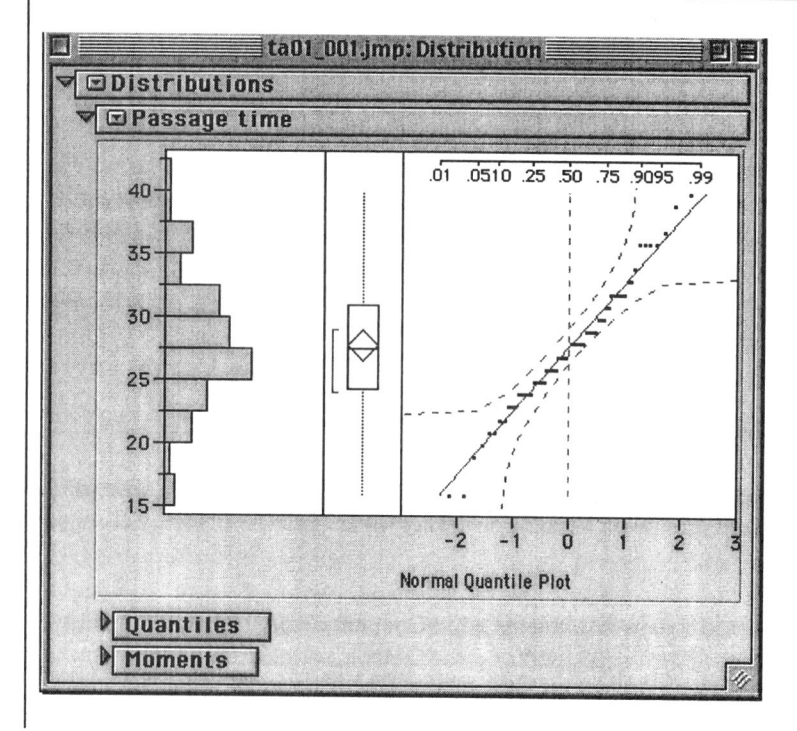

The 64 measurements closely follow a normal distribution so the use of a *t* confidence interval is justified. Let's obtain a series of confidence intervals (99%, 95%, and 80%) for the mean μ of the distribution from which these 64 measurements are a sample.

7. Click on the red triangle next to **Passage Time** and select **Confidence Interval** ⇨ **.99**.

Repeat this for a 95% confidence interval and for an 80% interval to obtain the following reports.

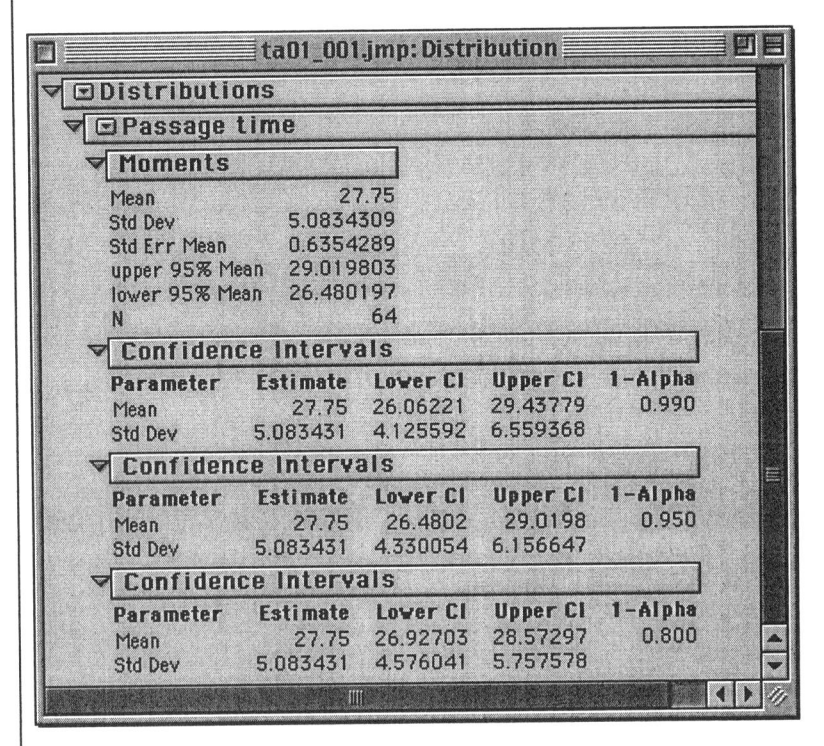

```
ta01_001.jmp: Distribution
Distributions
  Passage time
    Moments
      Mean          27.75
      Std Dev       5.0834309
      Std Err Mean  0.6354289
      upper 95% Mean 29.019803
      lower 95% Mean 26.480197
      N             64
    Confidence Intervals
      Parameter  Estimate  Lower CI  Upper CI  1-Alpha
      Mean       27.75     26.06221  29.43779  0.990
      Std Dev    5.083431  4.125592  6.559368
    Confidence Intervals
      Parameter  Estimate  Lower CI  Upper CI  1-Alpha
      Mean       27.75     26.4802   29.0198   0.950
      Std Dev    5.083431  4.330054  6.156647
    Confidence Intervals
      Parameter  Estimate  Lower CI  Upper CI  1-Alpha
      Mean       27.75     26.92703  28.57297  0.800
      Std Dev    5.083431  4.576041  5.757578
```

The 99% confidence interval for Newcomb's estimated passage time of light is (26.062, 29.438). If we are willing to be less confident (95% or 80%), the interval will be more precise; i.e., the interval and the margin of error will be smaller.

Remarks

- There was no need to request the 95% confidence interval. Since 95% is a frequently used confidence level, JMP displays the endpoints of the 95% interval for the mean in the **Moments** report.

- You can change the number of decimal places displayed for the mean and confidence intervals. Simply double-click on any of the values and a numeric format dialog panel will be displayed. For more details on this and other interesting modifications that you might wish to make to the appearance of these reports, see Section 0.6 in Chapter 0, "Enhancing Reports and Graphs".

- Notice the diamond in the boxplot. Its horizontal axis is located at the sample mean (27.75) and the vertical span represents the 95% confidence interval (26.5, 29.0). JMP calls this a *means diamond* a displays it when the mean and standard deviation are appropriate descriptors for a distribution. We encountered means diamonds in Section 2.1.2 in Chapter 2. We will use them later in this chapter again in Chapters 12 and 13 when we compare means of several distributions.

7.1.2 The One-Sample *t* Test

We also use the **Distribution** platform to perform one-sample *t* significance tests.

IPS Example 7.5 Passage time of light (cont'd.)

We wish to see whether Newcomb's result is significantly different from that obtained using the best modern equipment. Thus, we test

$$H_a: \mu \neq 33.02$$

If you have closed the data table **ta01_001.jmp** from the previous example, open it again and set the row state of the outliers, the 6th and 10th measurements, to **Exclude**.

1. Use the **Distribution** command to display the distribution of **Passage Time**.

2. Click on the red triangle next to the title bar for the **Passage Time** report and select **Test Mean**.

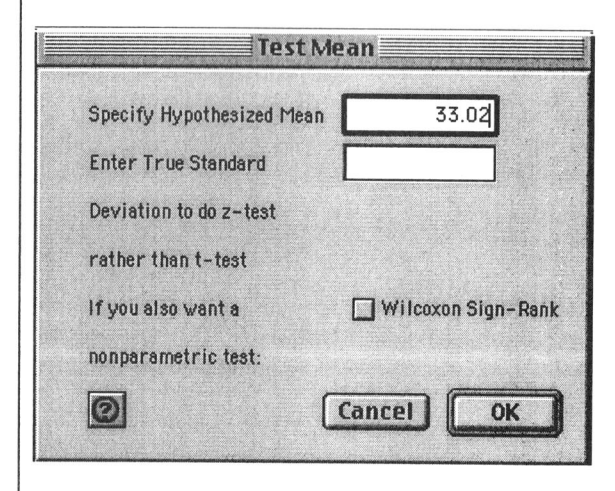

3. Type **33.02** in the `Specify Hypothesized Mean` field and select **OK**.

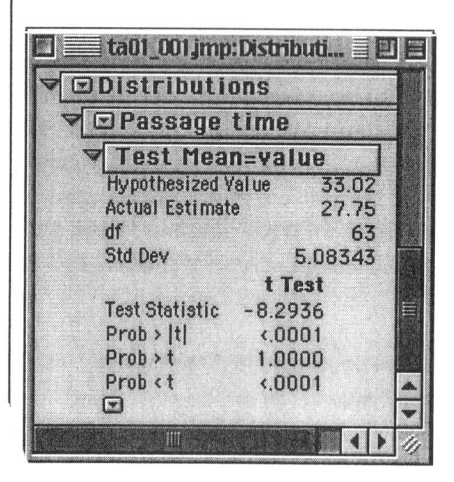

As was the case for the z test in Chapter 6, JMP calculates the value of the one-sample t test for the sample and provides three P-values, one for each of the three possible alternative hypotheses. We choose the one appropriate to the alternative hypothesis that we wish to test.

Alternative Hypothesis		P-value in JMP
$\mu \neq 33.02$	Two-tailed	Prob > |t|
$\mu > 33.02$	Upper tail	Prob > t
$\mu < 33.02$	Lower tail	Prob < t

Since our alternative hypothesis is that H_a: $\mu \neq 33.02$, we use the P-value for a two-sided test. Thus, we conclude that the mean of Newcomb's passage times, $\bar{x} = 27.75$, is lower than the value obtained by modern methods, $\mu = 33.02$ ($t = -8.2936$, $df = 63$, $P < 0.001$).

7.1.3 Matched Pairs t Procedures

The analysis of a matched pairs design is straightforward. A matched pairs study has two measurements on each individual and the objective is to compare the measurements by examining the difference between the measurements. To do this, we create a new column that contains the difference between the two measurements for each individual. Then, we analyze the difference using one-sample t procedures as above.

IPS Example 7.7 Immersion and understanding of spoken French

The NEH sponsored a study to determine whether 4 weeks of total immersion in French would improve the understanding of spoken French in high school teachers. Twenty French teachers were given an MLA listening test before and after 4 weeks of total immersion in French. To assess whether the immersion significantly improves understanding of spoken French, we test that the mean change in the test scores from before to after was positive. Thus, the alternative hypothesis

$$H_a: \mu > 0$$

where μ is the mean gain in score (i.e., the posttest score minus the pretest score).

This is a matched pairs study. Each teacher (individual) has two scores and we compare the scores by examining the difference between them. We create a new variable called **Gain** that holds the differences. The 20 differences form a single sample. We then use the **Test Mean** command to determine whether the mean gain is greater than zero.

1. Import the data file **ta07_001.txt** from the IPS CD-ROM and name the first three columns **Teacher**, **Pretest** and **Posttest**, respectively.

A fourth column, which represents the difference between the Posttest and the Pretest scores, is included in the data file. Since this will not ordinarily be the case, we will create a new variable called **Gain**, which is the **Posttest** score minus the **Pretest** score (see Section 0.3.3 in Chapter 0.)

2. Select **Cols** ⇨ **New Column**.

3. Name the column **Gain** and then select **New Property** ⇨ **Formula** to open the Formula Editor window.

Inference for Distributions

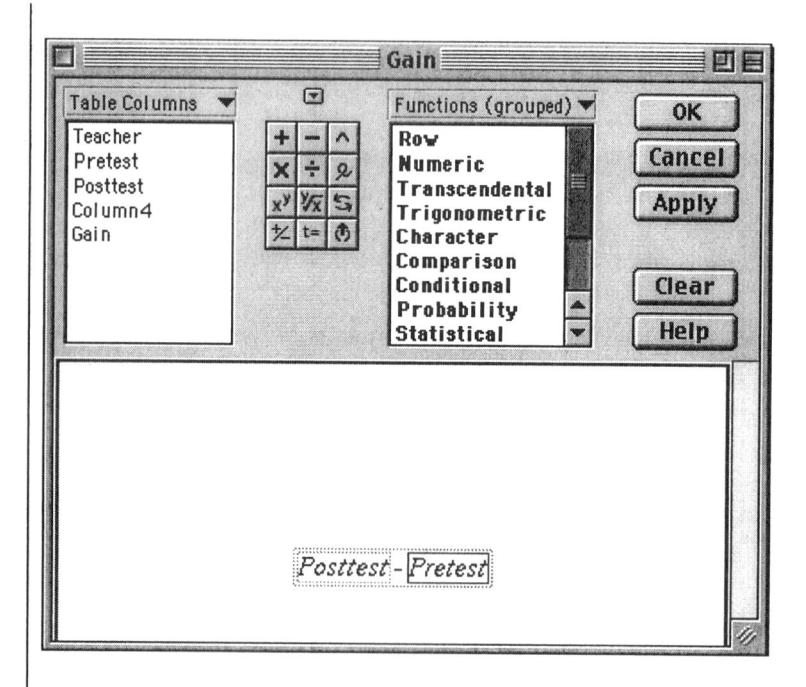

[Screenshot of JMP "Gain" formula dialog with formula: Posttest − Pretest]

    a. Select **Posttest** from the list of columns.
    b. Press **−** (minus) on the keypad and select **Pretest** from the list of columns.
    c. Press **OK** and **OK** again.

Check that the new variable **Gain** is indeed the **Posttest** score minus the **Pretest** score.

[Screenshot of ta07_001.JMP data table]

Row	Teacher	Pretest	Posttest	Column4	Gain
1	1	32	34	2	2
2	2	31	31	0	0
3	3	29	35	6	6
4	4	10	16	6	6
5	5	30	33	3	3
6	6	33	36	3	3
7	7	22	24	2	2
8	8	25	28	3	3
9	9	32	26	−6	−6

Let's calculate the mean pretest, the mean posttest score, and the mean improvement.

4. Select **Analyze** ⇨ **Distribution**.

5. Select **Pretest**, **Posttest**, and Gain from the list of columns and then press the Y, Columns and OK buttons.

☐

The mean Gain for the sample of 20 teachers was 2.5, which is the difference between the mean Posttest score 28.3 and the mean Pretest score 25.8. To test whether a mean improvement of 2.5 is statistically significant, we use the Test Mean command.

6. Press the same red triangle next to **Gain** and select **Test Mean**.

7. Type **0** in the `Specify Hypothesized Mean` field and select **OK**.

```
Gain
  Moments
    Mean              2.5
    Std Dev           2.8928223
    Std Err Mean      0.6468547
    upper 95% Mean    3.8538825
    lower 95% Mean    1.1461175
    N                 20
  Test Mean=value
    Hypothesized Value   0
    Actual Estimate      2.5
    df                   19
    Std Dev              2.89282
                    t Test
    Test Statistic   3.8649
    Prob > |t|       0.0010
    Prob > t         0.0005
    Prob < t         0.9995
```

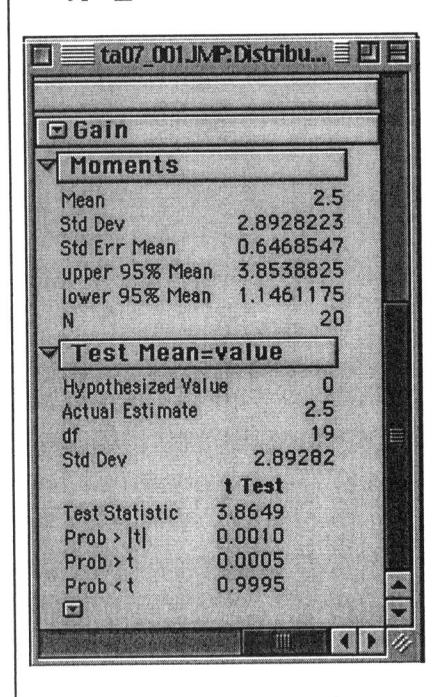

The mean improvement, $\bar{x} = 2.5$, was significant ($t = 3.8649$, $df = 19$, $P = 0.0005$). This conclusion is evident upon a close look at the boxplot for gain. The means diamond does not contain zero. A 95% confidence interval for the mean gain is (1.15, 3.85).

Remarks

- Look again at the JMP data table **ta07_001.jmp**. Each row represents a teacher's results. The scores are placed in two columns, not one, because each teacher has two scores. All matched pairs studies share a similar data table layout.

- Since a one-sample t test requires that the data be normally distributed, before submitting our results (actually, before performing the test), we would examine a normal quantile plot for gain. Press the red triangle next to the report on **Gain** and select **Normal Quantile Plot**.
  We notice an outlier (–6) and a staircase-like pattern, which indicates that only whole number scores were possible. For further discussion of the use of t procedures in the face of nonnormality, see the discussion on the robustness of the t procedures in the textbook.

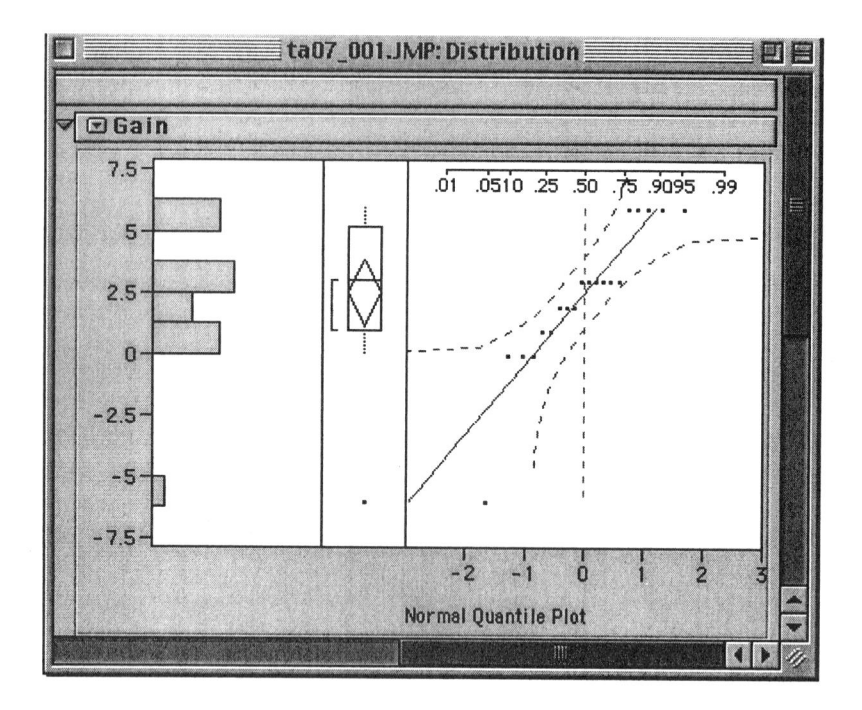

- JMP offers another approach to the analysis of matched pairs studies—the **Matched Pairs** command found on the Analyze menu. It provides the same output as above together with an interesting plot. To learn more about the command and plot, select **Analyze** ⇨ **Matched Pairs**. Then, press the **Help** button.

7.2 Comparing Two Means

Completely randomized designs are among the most common designs encountered in statistical practice. Two or more groups of individuals are compared with regard to a response variable. Unlike the matched pairs designs, there is no matching of the units in each group or sample. No individual belongs to more than one group and the responses in one group are independent of those in the other group. Hence, the data table for such a design consists of a column for the response variable and a column to identify the groups or treatments. The column that identifies the groups or treatments can be thought of as an explanatory variable or a factor. Then, comparing groups is equivalent to seeing if the response variable depends on the explanatory variable. To put it another way, we are interested in the relationship between the two variables.

We used the **Fit Y by X** analysis platform to look at relationships between two variables in Chapter 2. Side-by-side comparisons of the distributions of the response variable for each group were displayed with side-by-side boxplots and side-by-side means diamonds. To extend our conclusions to the populations from which the data are samples, we use the **Means/Anova/t-Test** command. Let's illustrate this for the study described in Example 7.14 of the textbook.

IPS Example 7.14 Directed reading activity and reading ability

An educator studied the effect of new directed reading activities on the reading ability of third graders. A class of 21 students took part in the new activities and another class of 23 followed the same curriculum without these activities. The Degree of Reading (DRP) test scores for the 44 students are found in the file **ta07_003.txt** on the IPS CD-ROM. The educator hoped to show that children who receive the new activity (group 1) will do better than those who do not (group 2); the alternative hypothesis to be tested is

H_a: $\mu_1 > \mu_2$

To perform the analysis in JMP, the data must be placed in a JMP data table in a certain way. There must be 44 rows, one for each student. There must be a column containing the DRP scores, the response variable, and a column identifying the group to which the student was assigned.

1. Import the file **ta07_003.txt** into JMP to see how the data is coded. Fortunately, the data file is in the correct format.

2. Name the four columns **Student**, **Group**, **Group code**, and **DRP score**, respectively.

Proceed as in Section 2.1.2 in Chapter 2 to study the relationship between the DRP scores and the groups.

3. Select **Analyze** ⇨ **Fit Y by X**.
   a. Select **DRP score** ⇨ **Y, Response**.
   b. Select **Group** ⇨ **X, Factor** and then select **OK**.

4. Click on the red triangle next to **Oneway Analysis of DRP score By Group** and select **Means/Anova/t-Test**.

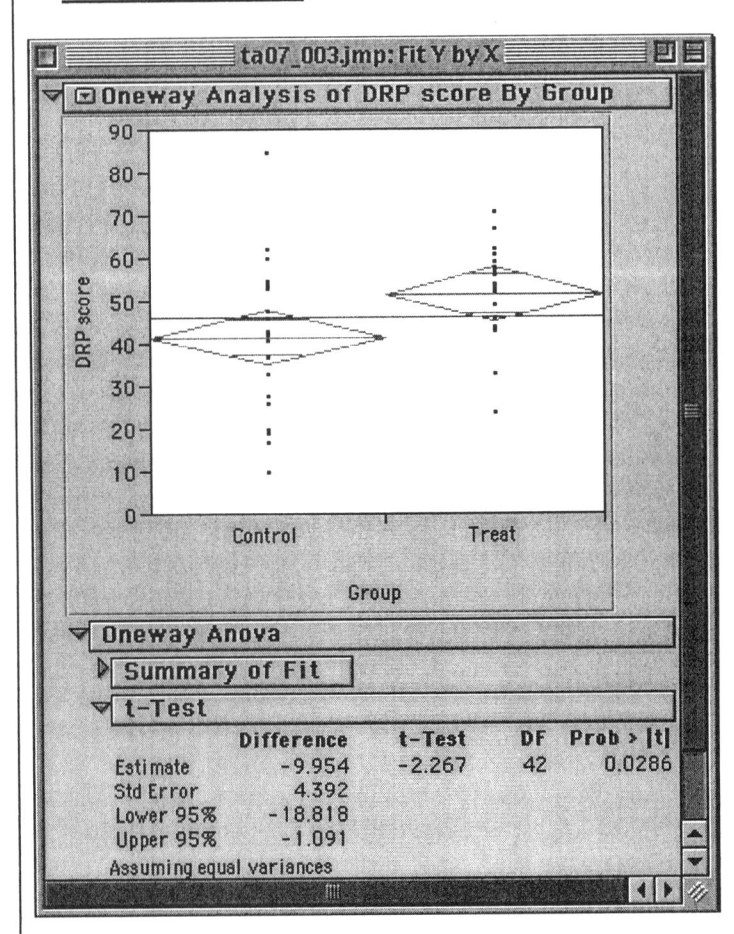

```
              ta07_003.jmp: Fit Y by X
  Oneway Analysis of DRP score By Group
  90
  80
  70
  60
  50
  40
  30
  20
  10
   0
         Control            Treat
                   Group
  Oneway Anova
  Summary of Fit
  t-Test
              Difference   t-Test    DF    Prob > |t|
  Estimate      -9.954     -2.267    42      0.0286
  Std Error      4.392
  Lower 95%    -18.818
  Upper 95%     -1.091
  Assuming equal variances
```

Inspect the side-by-side means diamonds. The vertical span of each diamond represents the 95% confidence interval estimate for the mean of each group. The horizontal lines above and below each group mean are called overlap marks. For groups with equal sample sizes, overlapping of these marks indicates that the two groups means are not significantly different at the .05 significance level (for a two-sided test). Because the diamonds assume that the groups have equal variances, we should only use the means diamonds as rough indicators.

Inspect the **t-Test** report. The pooled estimate of the variance is used to calculate the *t*-statistic and the confidence interval.

Estimate   difference between the two sample means (−9.954 points in this example)
Std. Error   estimated standard error of the difference between the two sample means using the pooled estimate of the variance
Lower 95%   lower bound on the 95% confidence interval for the difference between the means
Upper 95%   upper bound on the 95% confidence interval for the difference between the means

The 95% confidence interval estimate for the difference in mean DRP scores $\mu_1 - \mu_2$ between the groups of students is (−18.818, −1.091).

t-Test   value of the pooled two-sample *t* statistic
Prob > |t| *P*-value for a two-sided alternative hypothesis. For a one-sided alternative, the *P*-value is either half of this value or one minus half of it, depending on the direction of the one-sided alternative.

Thus, for our alternative hypothesis, the *P*-value is .5 × 0.0286 = 0.0143. We conclude that the data strongly suggests directed reading activity improves the DRP score (*t* = 2.267, *df* = 42, *P* = 0.014).

To obtain the two-sample *t* significance test *without assuming that the unknown population standard deviations are equal*,

5. Click on the red triangle again and select **Unequal Variances**.

We ignore the top part of the report for now. Look carefully at the results from the *Welch Anova* report.

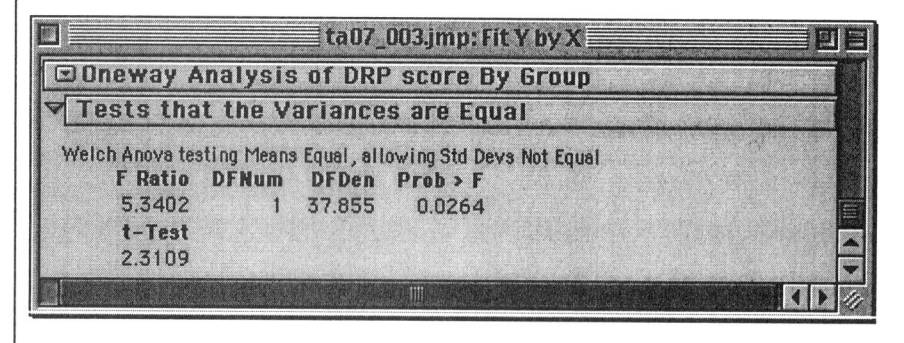

```
                    ta07_003.jmp: Fit Y by X
  Oneway Analysis of DRP score By Group
  Tests that the Variances are Equal
  Welch Anova testing Means Equal, allowing Std Devs Not Equal
    F Ratio   DFNum   DFDen    Prob > F
    5.3402      1     37.855    0.0264
    t-Test
    2.3109
```

t-Test   Two-sample *t* statistic
DFDen   Software approximation for the degrees of freedom
Prob > F   *P*-value for a two-sided alternative hypothesis. For a one-sided alternative, the *P*-value is either half of this value or one minus half of it, depending on the direction of the one-sided alternative.

Thus, for our example, we have *t* = 2.3109, *df* = 27.9, and *P* = 0.0132.

Our *P*-values and hence our conclusions have assumed that the data are normally distributed. To assess this, we look at normal quantile plots for the groups.

6. Click on the red triangle next to **Oneway Analysis of DRP score By Group** and select **Normal Quantile Plot** ⇨ **Plot Quantile by Actual** to obtain normal quantile plots for both groups.

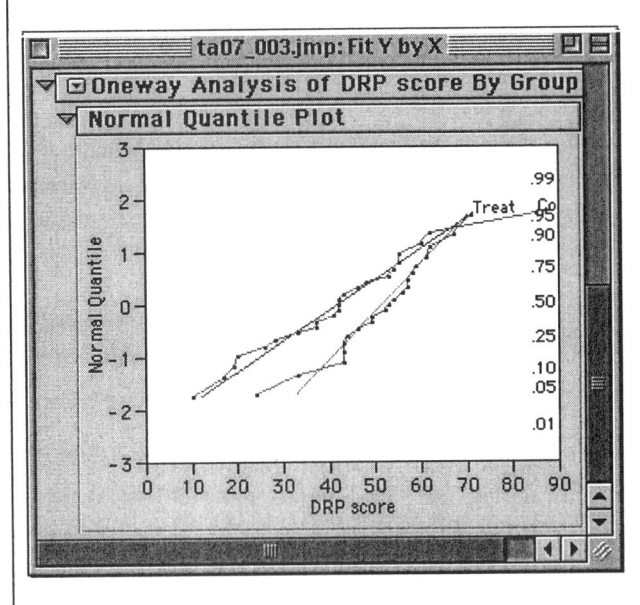

The plots confirm that the DRP scores for both groups are approximately normally distributed.

Remark

Sometimes the names of the treatments or groups are given numeric codes; for example, 0 and 1. To analyze the data, we must ensure that the *modeling type* of the group variable, or factor, is nominal. This is **very important** since JMP will display a scatterplot and offer tools for fitting curves when the explanatory variable has a continuous *modeling type* (the default for numeric data type). You can change the *modeling type* of a variable/column with the **Cols Info** command in the **Cols** menu. Alternatively, if the left table information panels (see Section 0.4 in Chapter 0) are open, you may select the icon next to the column name.

For the previous study, **Group code** is a numeric variable identifying the group to which each student belongs. The following figure illustrates using the icon on the left side panel to change the modeling type of **Group code**.

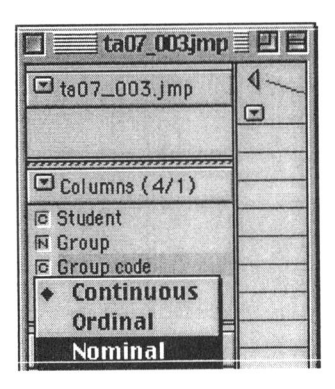

7.3 Testing for Unequal Variances

Pooled two-sample t procedures assume that the variances of both populations are equal. The **UnEqual Variances** command in the **Fit Y by X** platform in JMP provides several procedures for verifying that assumption.

IPS Example 7.14 Directed reading activity and DRP scores (cont'd.)

1. If the JMP data table from the last section is closed, import that file **ta07_003.txt** as described there.

2. Select **Analyze** ⇨ **Fit Y by X**.
    a. Select **DRP score** ⇨ **Y, Response**.
    b. Select **Group** ⇨ **X, Factor** and then select **OK**.

3. Click on the red triangle next to **Oneway Analysis of DRP score By Group** and select **UnEqual Variances**.

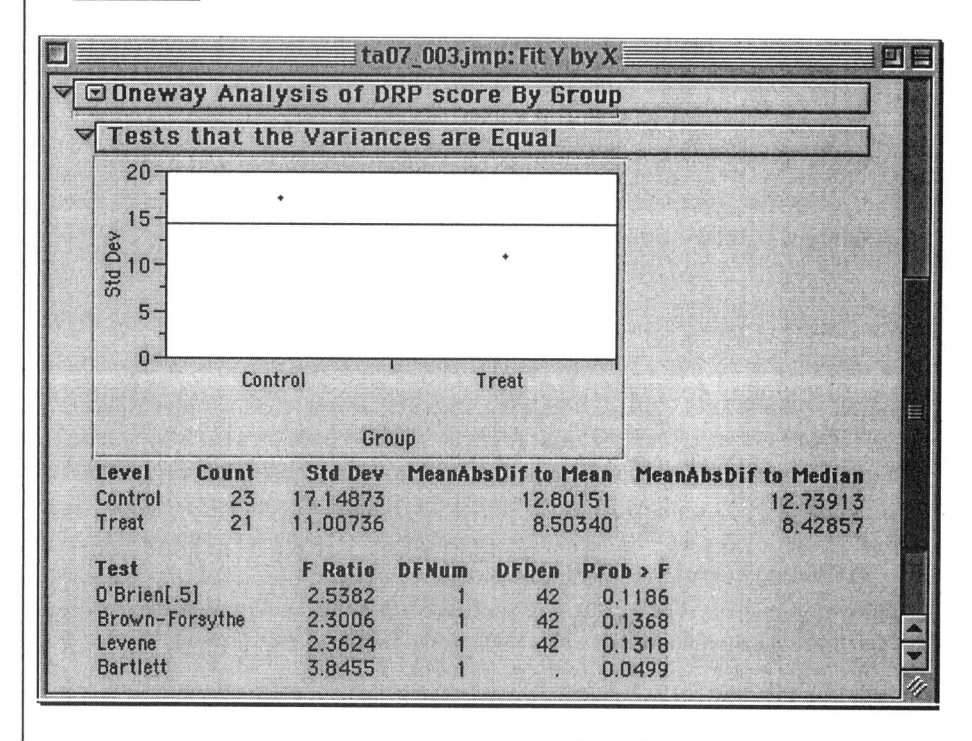

```
                    ta07_003.jmp: Fit Y by X
  Oneway Analysis of DRP score By Group
   Tests that the Variances are Equal

   20
   15
Std Dev
   10
    5
    0
        Control              Treat
                    Group

  Level    Count    Std Dev   MeanAbsDif to Mean   MeanAbsDif to Median
  Control    23    17.14873       12.80151              12.73913
  Treat      21    11.00736        8.50340               8.42857

  Test            F Ratio   DFNum   DFDen   Prob > F
  O'Brien[.5]      2.5382      1      42     0.1186
  Brown-Forsythe   2.3006      1      42     0.1368
  Levene           2.3624      1      42     0.1318
  Bartlett         3.8455      1              0.0499
```

Four different tests for unequal variances H_1: $\sigma_1^2 \neq \sigma_2^2$ are reported. To see an explanation of each test, select the [?] tool and click on any of the test names. The help window will open. P-values for the tests are given in the **Prob > F** column. In practice, if any of the tests are significant, you should **not** use the pooled estimate of the variance and pooled two-sample t procedures.

In this case, the P-value for Bartlett's test is less than 0.05. Thus, we would use the two-sample t test as found in the Welch Anova at the bottom of the report.

Remark

The tests for unequal variances described in this section and the assessment of normality of the response variable described in the previous section are also used when we compare more than two means in Chapters 12 and 13.

7.4 Summary

All graphs and statistical computations in this chapter are performed in the **Distribution** and **Fit Y by X** platforms of the **Analyze** menu.

Graph/Computation	Command
One-sample procedures	
Confidence intervals	**Distribution** ⇨ **Confidence Intervals**
Significance tests	**Distribution** ⇨ **Test Mean**
Matched pairs*	**Distribution** ⇨ **Confidence Intervals** and **Distribution** ⇨ **Test Mean**
Two-sample procedures**	**Fit Y by X** ⇨ **Means/Anova/t-Test**
Evaluating assumptions	
Normal quantile plots	**Fit Y by X** ⇨ **Normal Quantile Plots**
Equal variances	**Fit Y by X** ⇨ **Unequal Variances**

* The variable is the *difference* between the paired values.
** The *modeling type* of the explanatory variable must be nominal. The default modeling type of character variables is nominal, but the default modeling type of numeric variables is continuous.

7.5 Exercises

1. *t* procedures are robust against many forms of nonnormality with the exception of outliers and skewness. The NEH data in IPS Example 7.7 in Section 7.1.3 contains an outlier. To judge the effect of the outlier, rerun the *t* test without it. Can you still conclude that immersion has a significant effect?

2. In IPS Example 7.14, we used **Group** as the explanatory variable, or 'X, Factor' in the **Fit Y by X** dialog.
    a. Repeat the analysis, but replace the column **Group** with **Group code** in step 5. What happens in step 6?
    b. Repeat part a, but first change the modeling type of **Group code** to nominal. What happens in step 6 this time?

You should use JMP on any problem in the IPS text for which the raw data are provided. Here are some suggestions:

Section 7.1: 7.2, 3, 4, 9, 16, 17, 18, 19, 20, 23, 29, 30, 31, 32, 37, 39, 40, 41, 42

Section 7.2: 7.58, 59, 65, 66, 68, 69, 76, 77, 78

Section 7.3: 7.92, 94, 97

Chapter 7 Exercises: 7.103, 104, 105, 106, 111, 118, 120, 122, 123, 124, 131

Chapter 8

Inference for Proportions

This chapter is the first of two that deals with response variables that are categorical rather than quantitative. Initially, inference for one population is discussed. Then, inference for comparing the proportion in two populations is presented. The case of three or more populations and the more general question of whether two categorical variables are related are discussed in Chapter 9. All graphs and statistical computations in this chapter are performed in the second platform **Fit Y by X** of the **Analyze** menu.

8.1 Inference for a Population Proportion

JMP does not have commands for calculating confidence intervals or performing significance tests on a single population proportion.

8.2 Comparing Two Proportions

Just as comparing two means is equivalent to testing whether a categorical variable and a quantitative variable are related, comparing the proportion of successes in two groups is equivalent to testing whether two categorical variables are related. The groups form a categorical explanatory variable, or factor. The response variable is also categorical, taking the value "success" or "failure" (non-success) for each individual. To compare two proportions, JMP uses the same platform **Fit Y by X** that was used for comparing two population means. JMP recognizes the different modeling type of the response variable and automatically provides the appropriate test statistic and *P*-value for comparing two proportions.

IPS Examples 8.9 Binge drinking and gender

Are men and women college students equally likely to be frequent binge drinkers? A survey of over 17,000 students in U.S. four-year colleges collected information on drinking behavior. Here is the data summary.

Gender	Sample Size	Number of FBDs	Sample Proportion
Men	7180	1630	0.227
Women	9916	1684	0.170

The sample proportions of frequent binge drinkers (FBDs) are very different. We will perform a significance test to see whether this difference between the two samples is large enough to conclude that the proportion in the population of all male college students and the proportion in the population of all female college students are not equal. Therefore, we test the hypothesis

H_o: $p_M = p_F$ against the two-sided alternative H_a: $p_M \neq p_F$

where p_M and p_F denote the population proportion of frequent binge drinkers among all male and female students in U.S. colleges, respectively.

To perform the analysis in JMP, the data must be first placed in a JMP data table. Each of the 17,096 students in the study is either male or female and either a frequent binge drinker or not. Thus, we have a variable, **Gender**, with two categories, and a variable, **Frequent Binge Drinker**, with two categories, for a total of $2 \times 2 = 4$ classes. Each of the 17,096 men in the study falls into one of these four classes. Rather than entering 17,096 rows in the table, we use a column **Count** to summarize the number of individuals, students in this case, in each of the four gender-by-response categories.

1. Select **File** ⇨ **New** from the menu bar to create a new JMP data table.

2. Select **Cols** ⇨ **Add Multiple Columns** to accommodate the 3 variables.

3. Enter **3** after `How many columns to add` and press **OK**.

4. Change the names of the columns to `Gender`, `Frequent Binge Drinker`, and `Count`.

By default, columns contain numeric data. Use the **Column Info** command on the **Cols** menu to change the **Data Type** of the first two variables. (Be sure to select a column in the data table before using the Cols menu.) Notice that the *modeling type* automatically changes from "Continuous" to "Nominal."

5. Select **Rows** ⇨ **Add Rows** from the menu bar and enter **4**.

6. Fill in the data grid as below.

	Gender	Frequent Binge Drinker	Count
1	Male	Yes	1630
2	Male	No	5550
3	Female	Yes	1684
4	Female	No	8232

7. Select **File** ⇨ **Save** and name the data table **eg08_009.jmp**.

Examine the sample proportions of students who are frequent binge drinkers for each sex.

8. Select **Analyze** ⇨ **Fit Y by X**.
    a. Select **Frequent Binge Drinker** from the list of columns and press **Y, Response**.
    b. Select **Gender** and press **X, Factor**.
    c. Select **Count** and press **Freq** and **OK**.

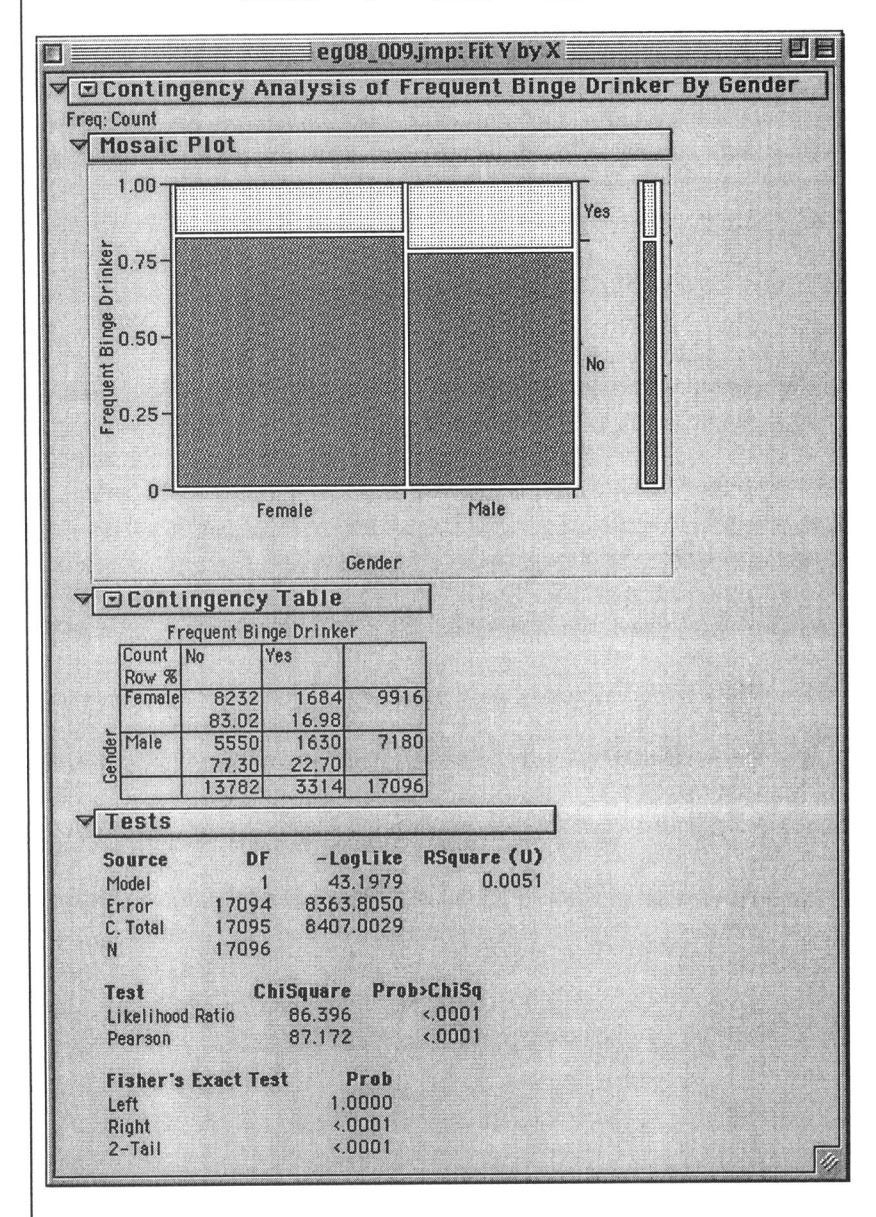

```
eg08_009.jmp: Fit Y by X
Contingency Analysis of Frequent Binge Drinker By Gender
Freq: Count
Mosaic Plot

Contingency Table
              Frequent Binge Drinker
Count   No      Yes
Row %
Female  8232    1684    9916
        83.02   16.98
Male    5550    1630    7180
        77.30   22.70
        13782   3314    17096

Tests
Source    DF      -LogLike    RSquare (U)
Model     1       43.1979     0.0051
Error     17094   8363.8050
C. Total  17095   8407.0029
N         17096

Test              ChiSquare   Prob>ChiSq
Likelihood Ratio  86.396      <.0001
Pearson           87.172      <.0001

Fisher's Exact Test   Prob
Left                  1.0000
Right                 <.0001
2-Tail                <.0001
```

The **Mosaic Plot** report displays three vertical stacked bar charts for the variable **Frequent Binge Drinker**. (Ignore the width of the bars.) The bar chart on the far right does not consider gender. The two adjacent ones on the left are for Female and Male college students, respectively. The green bar (Yes) for males is taller than that for females. To see the sample proportions more clearly,

9. Press the red triangle in the **Contingency Table** report and deselect both **Total %** and **Col %**.

16.98% of female college students are frequent binge drinkers versus 22.70% of male college students.

Now focus your attention on the middle of the **Tests** report. JMP does not provide the z statistic. Instead, it calculates the more general *Pearson chi-square test statistic*; this is discussed in Chapter 9 of the textbook. The *Pearson chi-square test* can be used to compare more than two population proportions and, for the case of two proportions, equals the square of the z statistic. The value of the *Pearson chi-square statistic* reported by JMP is 87.172, which is the square of the z test statistic, $z = 9.34$, given in the textbook. The *P*-value (labeled **Prob>ChiSq**) for a two-sided alternative is less than 0.0001. The difference is clearly statistically significant.

For a one-sided alternative, the *P*-value is either half of the **Prob>ChiSq** value or one minus half of it, depending on the direction of the one-sided alternative.

8.3 Summary

All graphs and statistical computations in this chapter are performed in the second platform **Fit Y by X** of the **Analyze** menu.

Activity	Command
Compare two proportions	**Analyze ⇨ Fit Y by X**

8.4 Exercises

1. Preferences for artificial Christmas trees. Exercise 8.41.

2. The health effects of drinking contaminated water. Exercise 8.47.

3. Chromosome abnormalities and criminality. Exercise 8.48.

4. Gender and summer employment. Exercises 8.49 and 8.52.

5. Gastric freezing as a treatment for ulcers. Exercise 8.77.

Chapter 9

Inference for Two-Way Tables

In Chapter 8, we compared two population proportions. In this chapter, we extend the comparison of proportions to more than two populations, to response variables with more than two categories, and to situations in which the sampling is from one population but two categorical variables are measured on each individual. *Two-way tables* provide models for each of these situations. Tools for displaying and describing the relationship between two categorical variables are presented first. Then, inference for extending conclusions to the population or populations that produced the data is studied.

Since analysis of two-way tables is equivalent to examining the relationship between two variables, all graphs and statistical computations in this chapter are performed in the second platform **Fit Y by X** of the **Analyze** menu. Creating a JMP data table that appropriately summarizes the information in the two-way table is critical.

9.1. Data Analysis for Two-Way Tables

JMP uses *mosaic plots* to display the relationship between two categorical variables. *Two-way tables* and *conditional distributions* describe the relationship and help identify striking deviations from that relationship.

IPS Example 9.12 Motivation for participation in sports

Do men and women participate in sports for the same reasons? A study was conducted at a large university on 67 randomly chosen male athletes and 67 randomly chosen female athletes. Each student was classified into one of four categories based on a questionnaire about their sports goals. Here are the data in the form of a two-way table with four rows and two columns.

	Gender	
Goal	Female	Male
HSC-HM	14	31
HSC-LM	7	18
LSC-HM	21	5
LSC-LM	25	13
Total	67	67

The four categories for sports goals were high social comparison-high mastery (HSC-HM), high social comparison-low mastery (HSC-LM), low social comparison-high mastery (LSC-HM), and low social comparison-low mastery (LSC-LM).

First, we create an appropriate JMP data table to summarize the two-way table. Then, we use the **Fit Y by X** platform to look at the relationship between gender and motivation in sports.

The entries in this two-way table are the numbers of students in each sports goal-by-gender category. Each of the two columns represents one gender and each of the four rows represents one of the four possible goals. Thus, to enter the data into a JMP data table, we need one variable **Goal** with 4 categories and one variable **Gender** with 2 categories for a total of 4 × 2 = 8 goal-by-gender classes. Instead of entering 134 rows, one for each student, we use 8 rows, one for each class, and include a column **Count** of the numbers of students in each class. Here is what the JMP data table should look like.

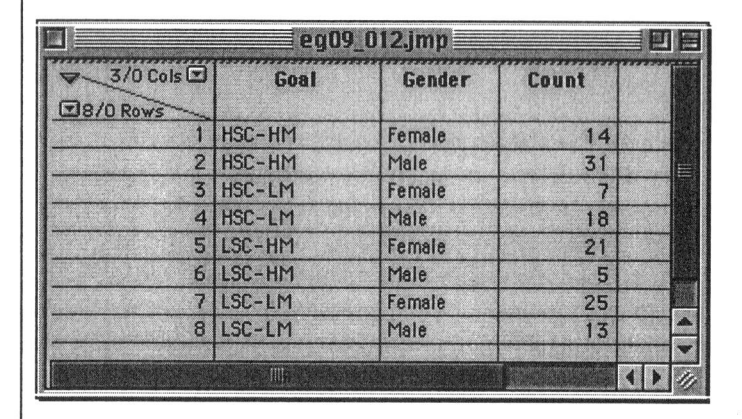

	Goal	Gender	Count
1	HSC-HM	Female	14
2	HSC-HM	Male	31
3	HSC-LM	Female	7
4	HSC-LM	Male	18
5	LSC-HM	Female	21
6	LSC-HM	Male	5
7	LSC-LM	Female	25
8	LSC-LM	Male	13

1. Select **File** ⇨ **New** from the menu bar.

2. Select **Cols** ⇨ **Add Multiple Columns** to accommodate the 3 variables.

3. Enter **3** after `How many columns to add` and press **OK**.

4. Change the names of the columns to `Goal`, `Gender`, and `Count`.

Use the **Column Info** command on the **Cols** menu to change the `Data Type` of the first two variables to "Character." (Be sure to select a column in the data table before using the **Cols** menu.) Notice that the `modeling type` automatically changes from "Continuous" to "Nominal." Here is the dialog for **Goal**.

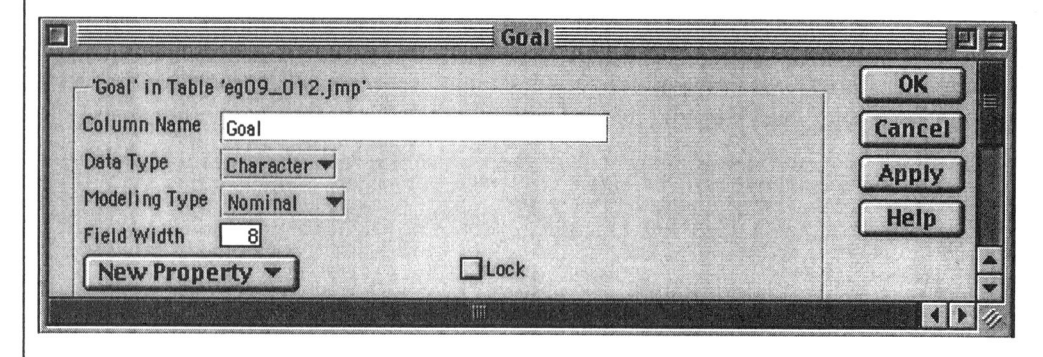

5. Select **Rows** ⇨ **Add Rows** from the menu bar and enter **8**.

6. Fill in the data grid as above.

7. Select **File** ⇨ **Save** and name the data table **eg09_012.jmp**.

Joint and Marginal Distributions

The joint and marginal distributions of **Goal** and **Gender** can be calculated and displayed using the **Fit Y by X** platform. The difference from previous uses of Fit Y by X is that we need to specify the frequency of each row, or goal-by-gender category. To create a distribution table with the same orientation as in the textbook, we will invert the roles of *Y* and *X*.

8. Select **Analyze** ⇨ **Fit Y by X**.

9. Select **Gender** and press **Y, Response**.

10. Select **Goal** and press **X, Factor**.

11. Select **Count** and press **Freq** and **OK**.

12. Press the red triangle in the **Contingency Table** report and deselect **Col %** and **Row %**.

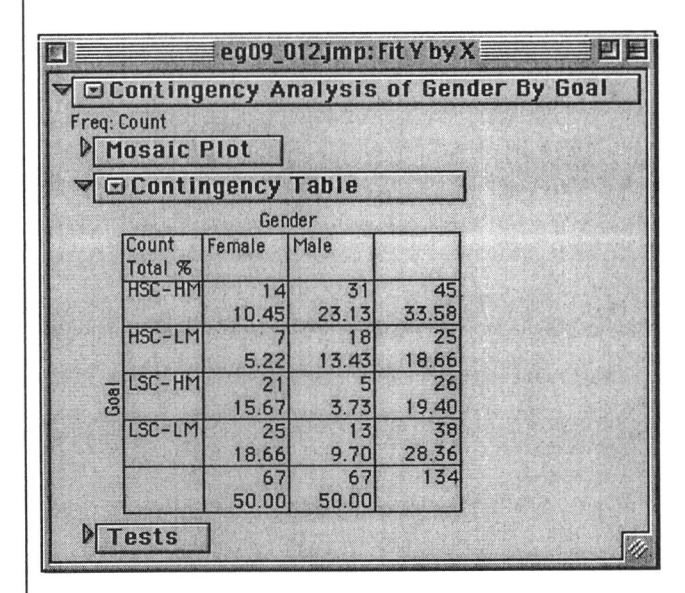

```
eg09_012.jmp: Fit Y by X
Contingency Analysis of Gender By Goal
Freq: Count
  Mosaic Plot
  Contingency Table
                    Gender
      Count    Female   Male
      Total %
      HSC-HM       14      31      45
                10.45   23.13   33.58
      HSC-LM        7      18      25
                 5.22   13.43   18.66
Goal  LSC-HM       21       5      26
                15.67    3.73   19.40
      LSC-LM       25      13      38
                18.66    9.70   28.36
                   67      67     134
                50.00   50.00
  Tests
```

The resulting report contains both counts and percents. The *joint* proportion of students who are female and in the high social comparison-high mastery group (HSC-HM) is 14 divided by 134, or 10.45%. The *marginal distribution* of sports goal is found in the right margin of the table. For example, 33.58% of all students in the study were in the HSC-HM group. The *marginal distribution* of gender is in the bottom margin of the table.

To study the effect of gender on the distribution of sports goal, we look at the *conditional distribution* of **Goal** for each **Gender**.

Conditional Distributions

13. Deselect **Count** and **Total %** in the **Contingency Table** report (use the red triangle menu) and select **Col %**.

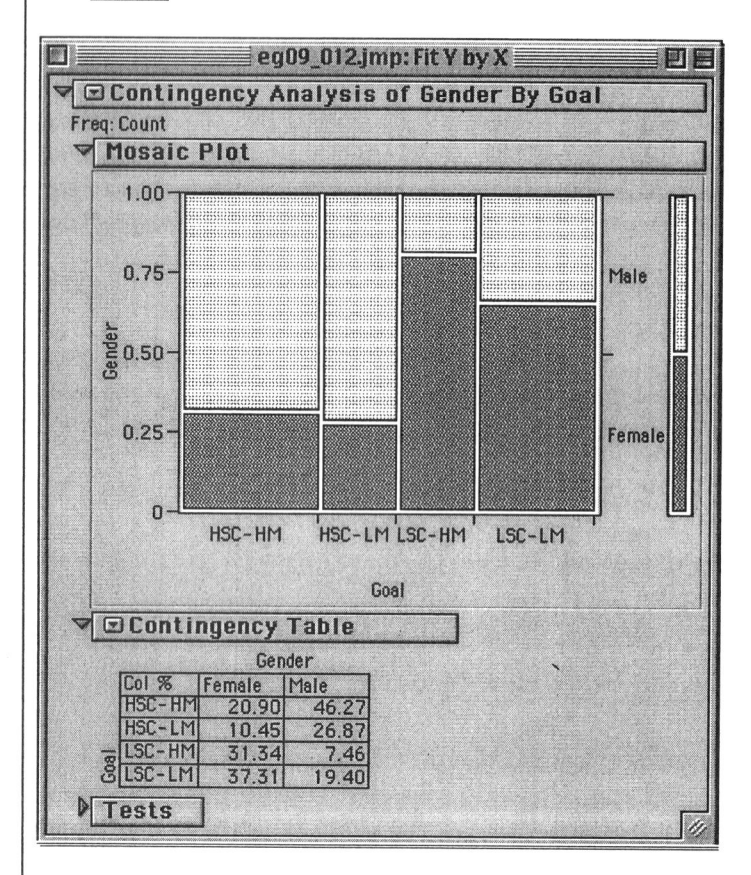

[Screenshot of JMP Contingency Analysis of Gender By Goal, showing Mosaic Plot and Contingency Table]

Col %	Female	Male
HSC-HM	20.90	46.27
HSC-LM	10.45	26.87
LSC-HM	31.34	7.46
LSC-LM	37.31	19.40

Both columns in the **Contingency Table** report add to 100%, one for females and one for males. Each column represents a *conditional distribution* of **Goal**. The distributions are not the same. If gender and sports goals are not related, then the percent of males and females would be the same for each goal. The Mosaic Plot clearly shows that this is not the case. The heights of the green (Male) and red (Female) bars within each category of sports goal are not the same.

To display the *conditional distributions* as stacked bar charts, the roles of the **Goal** and **Gender** variables must be assigned in the normal way with **Goal** as the response variable and **Gender** as the explanatory variable.

14. Select **Analyze** ⇨ **Fit Y by X**.
    a. Select **Goal** and press **Y, Response**.
    b. Select **Gender** and press **X, Factor**.
    c. Select **Count** and press **Freq** and **OK**.

Three stacked bar charts are displayed. The one on the far right displays the marginal distribution of **Goal**. The other two display the *conditional distributions* of **Goal** for each gender. Using the left scale, notice that the height of the orange and blue bars for the females is about .70 (1.00 –0.30). About 70% of female students have low social comparison goals (either LSC-LM or LSC-HM). Only about 25% of the males have low social comparison goals. For this data, male and female athletes clearly have different distributions of sports goals. In Section 9.2, a statistical test will tell us whether the difference can be attributed to chance or to a difference between male and female attitudes in general.

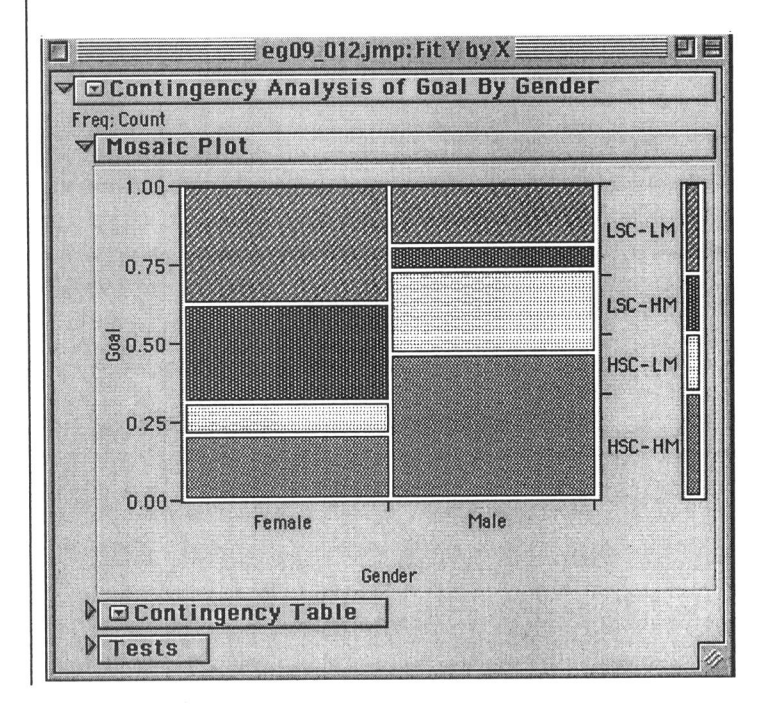

Remark

If the values of a categorical variable are numeric, you must change the modeling type to "Nominal."

Simpson's Paradox

We can use JMP to evaluate the effects of a lurking variable on the relationship between two categorical variables.

IPS Examples 9.9 and 9.10 Which hospital is safer?

Suppose that you wish to compare the survival rates of patients who have recently undergone surgery at two hospitals. Data on the survival of patients who underwent surgery are available along with the general condition of the patient before surgery. We first look at the relation of patient outcome and hospital without considering the condition of the patient. Then, we will examine the relation separately for patients in different conditions.

1. Select **File** ⇨ **New** and create four columns named **Outcome**, **Hospital**, **Condition**, and **Count**.

2. Enter the data as follows and save the JMP data table as **eg09_010.jmp**.

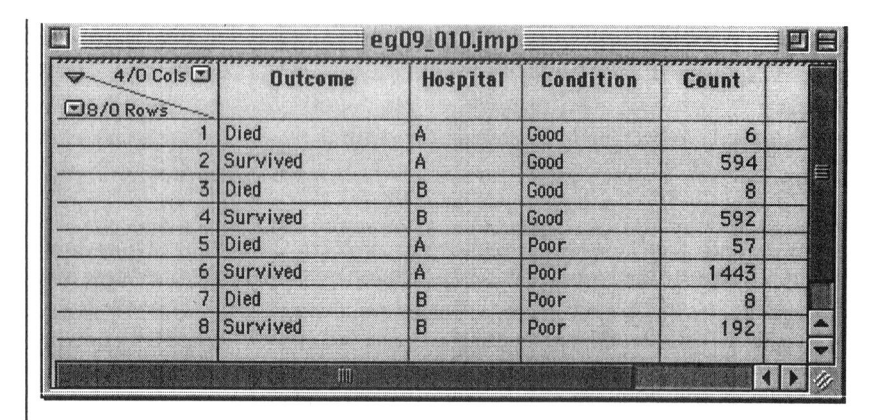

3. Select **Analyze** ⇨ **Fit Y by X**. To obtain the same orientation of the two-way table as in the textbook, invert the roles of **Hospital** and **Outcome**.
    a. Select **Hospital** and press **Y, Response**.
    b. Select **Outcome** and press **X, Factor**.
    c. Select **Count** and press **Freq** and **OK**.

4. Deselect **Total %** and **Row %** in the **Contingency Table** report to display the conditional distribution of **Outcome** for each hospital.

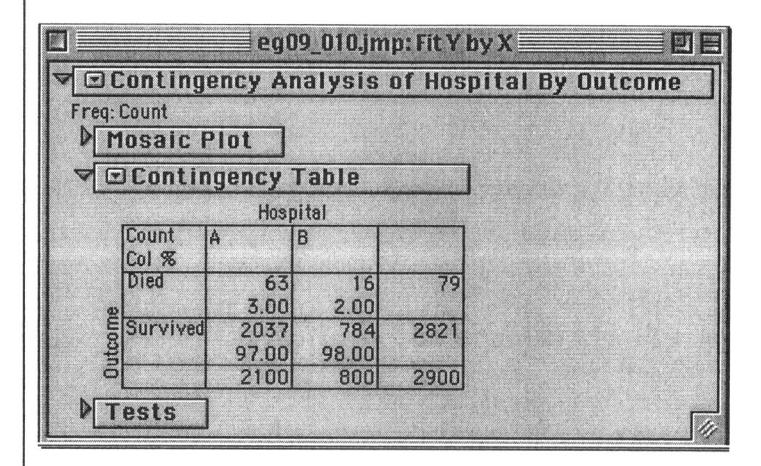

The table indicates that Hospital B is better—its death rate is lower (2%) than the death rate for Hospital A (3%). Now let's look at the effect of the patient's **Condition** on the relationship between **Outcome** and **Hospital**. To do this in JMP, we simply use the **By** option on the **Fit Y by X** platform.

5. Select **Analyze** ⇨ **Fit Y by X**.
    a. Select **Hospital** and press **Y, Response**.
    b. Select **Outcome** and press **X, Factor**.
    c. Select **Count** and press **Freq**.
    d. Select **Condition** and press **By** and **OK**.

6. Deselect **Total %** and **Row %** in each **Contingency Table** report.

The death rate is slightly lower for Hospital A, both among patients in good condition and among patients in poor condition. Thus, Hospital A is the safer hospital. Clearly, it is not a good idea to ignore a lurking variable when considering the relationship between two variables.

```
┌─────────────────────────────────────────────────┐
│ ▽        eg09_010.jmp: Contingency              │
│ ▽ Condition=Good                                │
│   ▽ ⊟ Contingency Analysis of Hospital By Outcome│
│     Freq: Count                                 │
│     ▷ Mosaic Plot                               │
│     ▽ ⊟ Contingency Table                       │
│              Hospital                           │
│        Count  A       B                         │
│        Col %                                    │
│        Died      6      8     14                │
│                1.00   1.33                      │
│        Survived 594   592   1186                │
│               99.00  98.67                      │
│                 600    600   1200               │
│     ▷ Tests                                     │
│ ▽ Condition=Poor                                │
│   ▽ ⊟ Contingency Analysis of Hospital By Outcome│
│     Freq: Count                                 │
│     ▷ Mosaic Plot                               │
│     ▽ ⊟ Contingency Table                       │
│              Hospital                           │
│        Count  A       B                         │
│        Col %                                    │
│        Died     57      8     65                │
│                3.80   4.00                      │
│        Survived 1443  192   1635                │
│               96.20  96.00                      │
│                1500    200   1700               │
│     ▷ Tests                                     │
└─────────────────────────────────────────────────┘
```

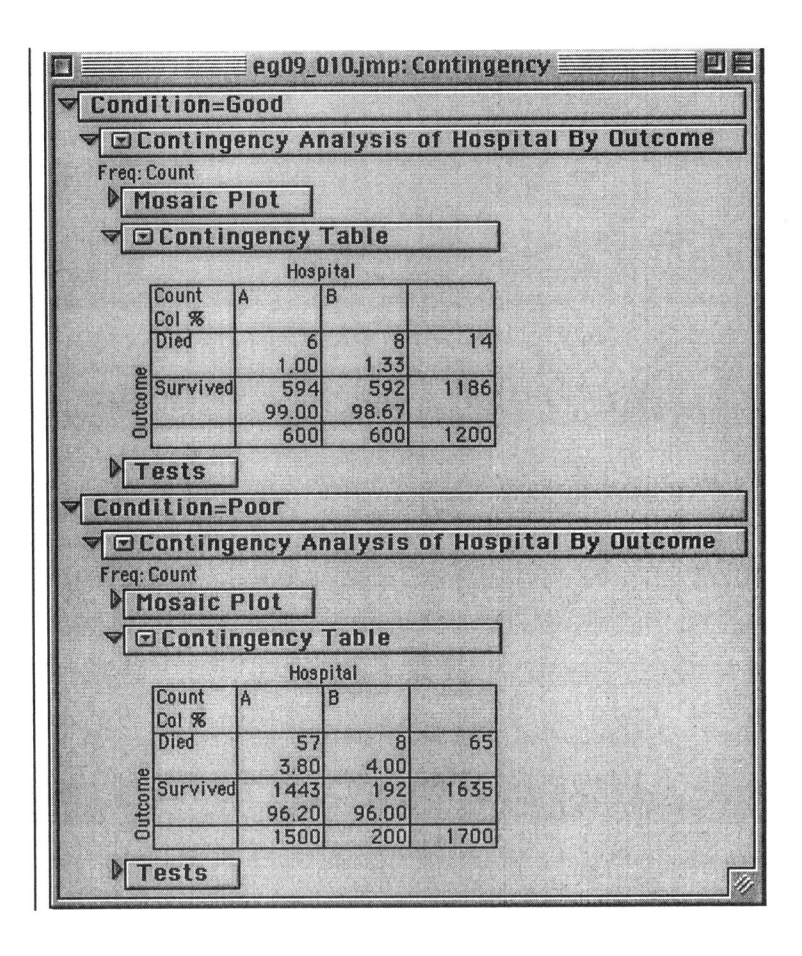

Remark

By default, JMP arranges the values of a categorical variable in alphabetical order. To change this ordering, use the column command **Validation**. We illustrate this by rearranging the three categories of socioeconomic status in Example 9.17 in the textbook.

IPS Example 9.17 Socioeconomic status and smoking habits

In a study of heart disease, researchers classified 356 volunteer subjects according to their socioeconomic status (SES) and their smoking habits.

1. Create a new JMP data table with the variables **Smoking**, **SES**, and **Count** and enter the values from the following figure.

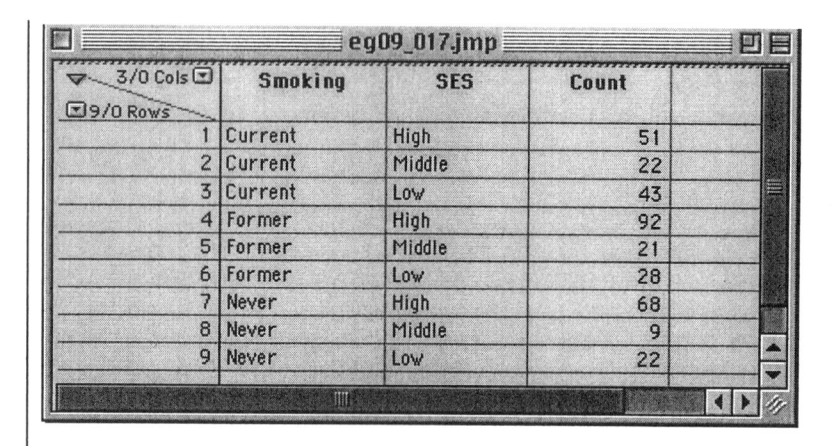

[Screenshot of JMP data table eg09_017.jmp with columns Smoking, SES, Count:]

	Smoking	SES	Count
1	Current	High	51
2	Current	Middle	22
3	Current	Low	43
4	Former	High	92
5	Former	Middle	21
6	Former	Low	28
7	Never	High	68
8	Never	Middle	9
9	Never	Low	22

Use the **Fit Y by X** platform to analyze the relationship between **Smoking** and **SES**.

2. Select **Analyze** ⇨ **Fit Y by X**.
    a. Select **SES** and press **Y, Response**.
    b. Select **Smoking** and press **X, Factor**.
    c. Select **Count** and press **Freq** and **OK**.

3. Deselect **Total %**, **Col %**, and **Row %** in the **Contingency Table** report.

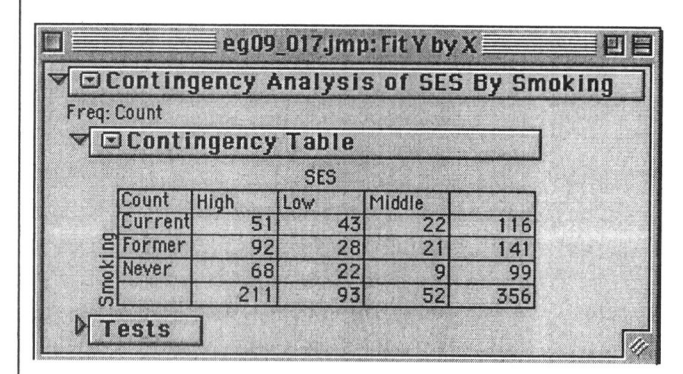

[Screenshot of Contingency Analysis of SES By Smoking, Freq: Count]

Contingency Table

Count	High	Low	Middle	
Current	51	43	22	116
Former	92	28	21	141
Never	68	22	9	99
	211	93	52	356

JMP prints the **SES** categories in alphabetical order: High, Low, Middle. We wish the more natural order given in the textbook: High, Middle, Low. Use the column command **Validation** to rearrange the categories.

4. Select the column **SES** in the data table.

5. Select **Cols** ⇨ **Validation** ⇨ **List Checking**.

A dialog box with the levels of **SES** is displayed. Select a level and press either **Move Up** or **Move Down**. The new order will be used in mosaic plots and contingency tables.

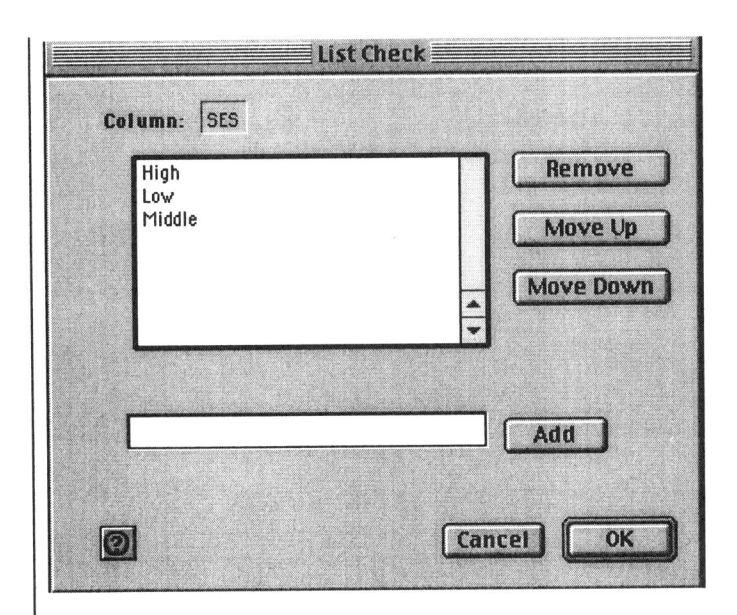

6. Select the category **Middle,** press **Move Up** to place "Middle" above "Low," and press **OK**.

Now rerun the **Fit Y by X** command. Bring the **eg09_017.jmp: Fit Y by X** window forward.

7. Press the top red triangle on the **Contingency Analysis of SES By Smoking** report and select **Script ⇨ Redo Analysis**.

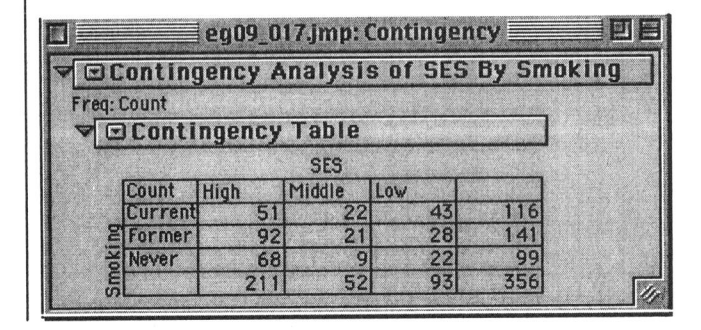

9.2 Inference for Two-Way Tables

The *Pearson chi-square statistic* allows conclusions about the data in hand to be extended to the population or populations that produced the data. The statistic together with an associated *P*-value is presented in the **Tests** report.

IPS Example 9.12 Motivation for participation in sports (cont'd.)

Do men and women participate in sports for the same reasons? A study was conducted at a large university on 67 randomly chosen male athletes and 67 randomly chosen female athletes. Each student was classified into one of four categories based on a questionnaire about their sports goals. We wish to test

H_o: There is no association between gender and sports goals against the alternative.

H_a: Gender and sports goals are related.

Open the JMP data table **eg09_012.jmp** created in Section 9.1 and use the **Fit Y by X** command.

1. **File** ⇨ **Open**.

2. Select **Analyze** ⇨ **Fit Y by X**.
   a. Select **Goal** and press **Y, Response**.
   b. Select **Gender** and press **X, Factor**.
   c. Select **Count** and press **Freq** and **OK**.

In Section 9.1, we concluded that the conditional distributions of sports goals for male and female athletes *in the sample* were different. We focus on the **Tests** report to determine whether that difference can be attributed to chance or to a difference between male and female attitudes in general.

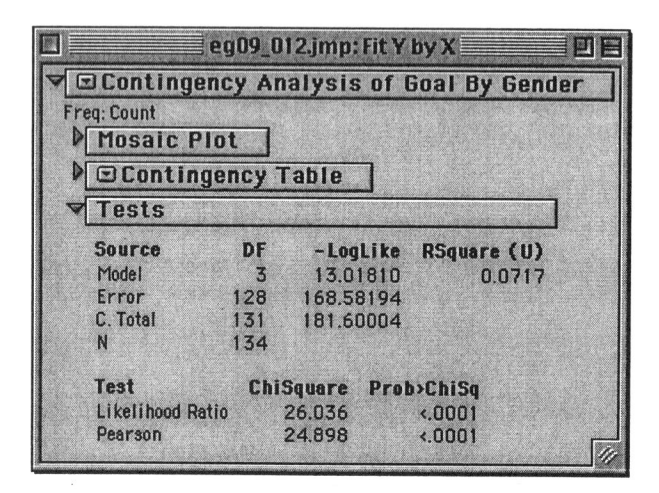

```
eg09_012.jmp: Fit Y by X
Contingency Analysis of Goal By Gender
Freq: Count
▶ Mosaic Plot
▶ Contingency Table
▼ Tests

Source      DF    -LogLike    RSquare (U)
Model        3    13.01810       0.0717
Error      128   168.58194
C. Total   131   181.60004
N          134

Test            ChiSquare   Prob>ChiSq
Likelihood Ratio   26.036      <.0001
Pearson            24.898      <.0001
```

The value of the *Pearson chi-square statistic* is 24.898 and the *P*-value (labeled **Prob>ChiSq**) is less than 0.0001. Since the *P*-value is small, the data indeed provide strong evidence that there are differences between genders in their reason for participation in sports. The mosaic plots in Section 9.1 clearly showed that males are more likely to be motivated by social comparison goals while females are more likely to be motivated by mastery goals.

Remark

JMP can display the expected counts for all cells in the table. Select **Expected** from the red triangle menu on the **Contingency Table** report. The expected counts are labeled "Expected" and are the last entry in each cell. For example, the expected number of HSC-HM females is 22.5.

Contingency Analysis of Gender By Goal

Freq: Count

Contingency Table

Count Total % Col % Row % Expected	Female	Male	
HSC-HM	14 10.45 20.90 31.11 22.5	31 23.13 46.27 68.89 22.5	45 33.58
HSC-LM	7 5.22 10.45 28.00 12.5	18 13.43 26.87 72.00 12.5	25 18.66
LSC-HM	21 15.67 31.34 80.77 13	5 3.73 7.46 19.23 13	26 19.40
LSC-LM	25 18.66 37.31 65.79 19	13 9.70 19.40 34.21 19	38 28.36
	67 50.00	67 50.00	134

9.3 Summary

All graphs and statistical computations in this chapter are performed in the second platform **Fit Y by X** of the **Analyze** menu. The modeling type of both variables must be "Nominal." To change the order of the categories of a variable, use the **Validation** command on the **Cols** menu.

Activity	Command
Inference for two-way tables	**Analyze** ⇨ **Fit Model** ⇨ ... ⇨ **Freq**
Simpson's paradox	**Fit Model** ⇨ ... ⇨ **By**
Reordering categories	**Cols** ⇨ **Validation** ⇨ **List Checking**

9.4 Exercises

1. Motivation for sports participation. Example 9.12. Use the **Distribution** platform to obtain the distributions of **Goal** and **Gender**. Compare these with the marginal distributions obtained in Section 9.1.

You should use JMP to solve any problem in this chapter that requires computation. For example,

2. Age and education. Exercises 9.1 through 9.3.

2. Treating cocaine addiction. Exercises 9.13 and 9.27.

3. Admission decisions and gender. Exercises 9.17 and 9.18.

4. Ulcers and gastric freezing. Exercise 9.43.

Chapter 10

Simple Linear Regression

In Section 2.3 of Chapter 2, we modeled the relationship between two quantitative variables as a straight line using least-squares regression. Now, we construct confidence intervals and perform significance tests for the unknown population regression parameters (the slope and the intercept) and the predicted values.

10.1 Simple Linear Regression

The first steps in the analysis of the relationship between two *quantitative* variables are to fit the model and to check the assumptions of the model using the residuals. We use the same JMP platform—**Fit Y by X**— that we used in Section 2.3 in Chapter 2.

10.1.1 Fitting the Model and Examining the Residuals

IPS Example 10.1 Body density and skinfold thickness

Body fat content is found in practice by measuring body density. High fat content corresponds to low body density. Body density is difficult to measure directly. For this reason, scientists have sought variables that are easier to measure and that can be used to predict body density. Research suggests that skinfold thickness can accurately predict body density. Skinfold thickness is measured at four body locations and summed. It is conjectured that the relation between body density and the log of the sum of the skinfold thickness measures is linear. Data on 92 males aged 20 to 29 was collected to test this conjecture.

1. Import the data from the text file **eg10_001.txt** on the IPS CD-ROM and name the variables **Person, Log Skin Thickness,** and **Body Density,** respectively.

Proceed as you did in Chapter 2 to fit the least-squares regression line for predicting body density from the logarithm of the sum of the skinfold thickness measures.

2. Select **Analyze** ⇨ **Fit Y by X**.

a. Select **Body Density** from the list of columns and press **Y, Response**.
b. Select **Log Skin Thickness** and press **X, Factor** and **OK**.
c. Select **Fit Line** from the red triangle menu.

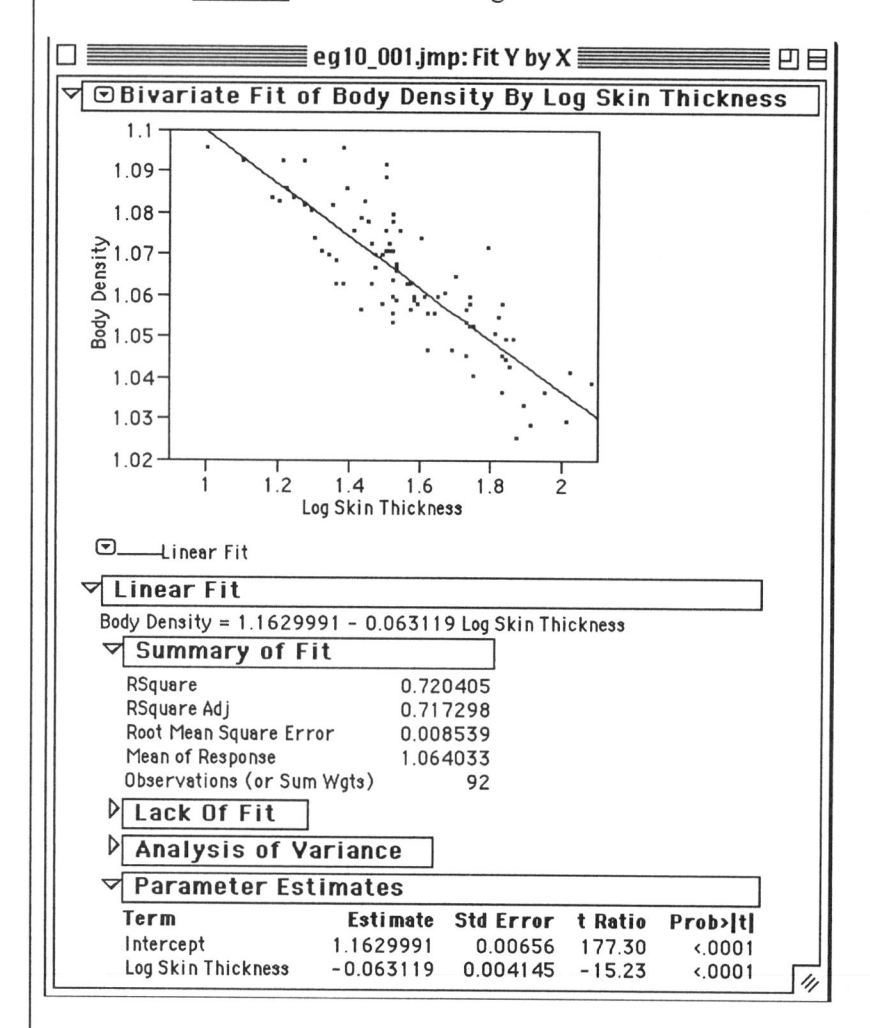

```
┌─────────────── eg10_001.jmp: Fit Y by X ───────────────┐
│ Bivariate Fit of Body Density By Log Skin Thickness    │
│                                                         │
│   [scatter plot with linear fit]                       │
│                                                         │
│   Linear Fit                                           │
│   Body Density = 1.1629991 - 0.063119 Log Skin Thickness│
│   Summary of Fit                                       │
│     RSquare                        0.720405            │
│     RSquare Adj                    0.717298            │
│     Root Mean Square Error         0.008539            │
│     Mean of Response               1.064033            │
│     Observations (or Sum Wgts)           92            │
│   Lack Of Fit                                          │
│   Analysis of Variance                                 │
│   Parameter Estimates                                  │
│     Term              Estimate  Std Error  t Ratio  Prob>|t|│
│     Intercept         1.1629991   0.00656   177.30   <.0001│
│     Log Skin Thickness -0.063119  0.004145  -15.23   <.0001│
└─────────────────────────────────────────────────────────┘
```

Inspect the **Linear Fit** report.

At the top, we find that the least-squares line is **Body Density**^ = 1.16 − 0.0631 **Log Skin Thickness**. In the **Summary of Fit** section, we find that r^2 (labeled **RSquare**) equals 0.7204 and that the *standard error about the line* (labeled **Root Mean Square Error**) equals $s = 0.00854$.

Before doing inference, we must check the required regression model assumptions. Checking assumptions involves examining the *residuals*. The JMP commands **Plot Residuals** and **Save Residuals** help us to do that.

Plot the Residuals Versus the Explanatory Variable

To plot the residuals of the 92 body density measurements versus the explanatory variable **Log Skin Thickness**, use the **Plot Residuals** command as we did in Section 2.4 in Chapter 2.

3. Select **Plot Residuals** from the red triangle menu next to **Linear Fit**.

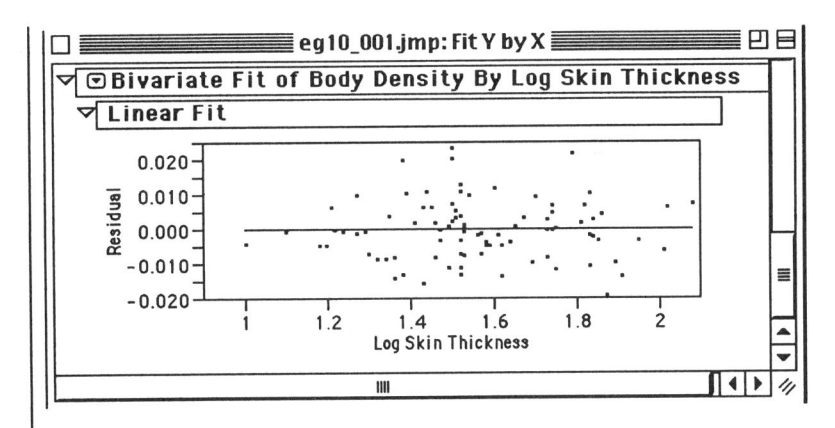

The plot appears at the bottom of the **Linear Fit** report. The residuals are on the vertical axis with a red line at zero. The scatter about the line is random with no unusual patterns.

Assessing Normality: Normal Quantile Plots

We save the residuals to the JMP data table and use **Normal Quantile Plot** command in the **Distribution** platform (Section 1.3 in Chapter 1) to check for signs of nonnormality.

4. Select **Save Residuals** from the red triangle menu for the **Linear Fit** report.

5. Select **Analyze** ⇨ **Distribution**.
   a. Select **Residuals Body Density** and press **Y, Columns** and **OK**.
   b. Press the red triangle for the **Residuals Body Density** report and select **Normal Quantile Plot**.

There are no signs of strong nonnormality. Observations 61 and 70 appear to be mild outliers. Further analysis (Section 2.4 in Chapter 2) indicates that they are not very influential and will have little effect on inference.

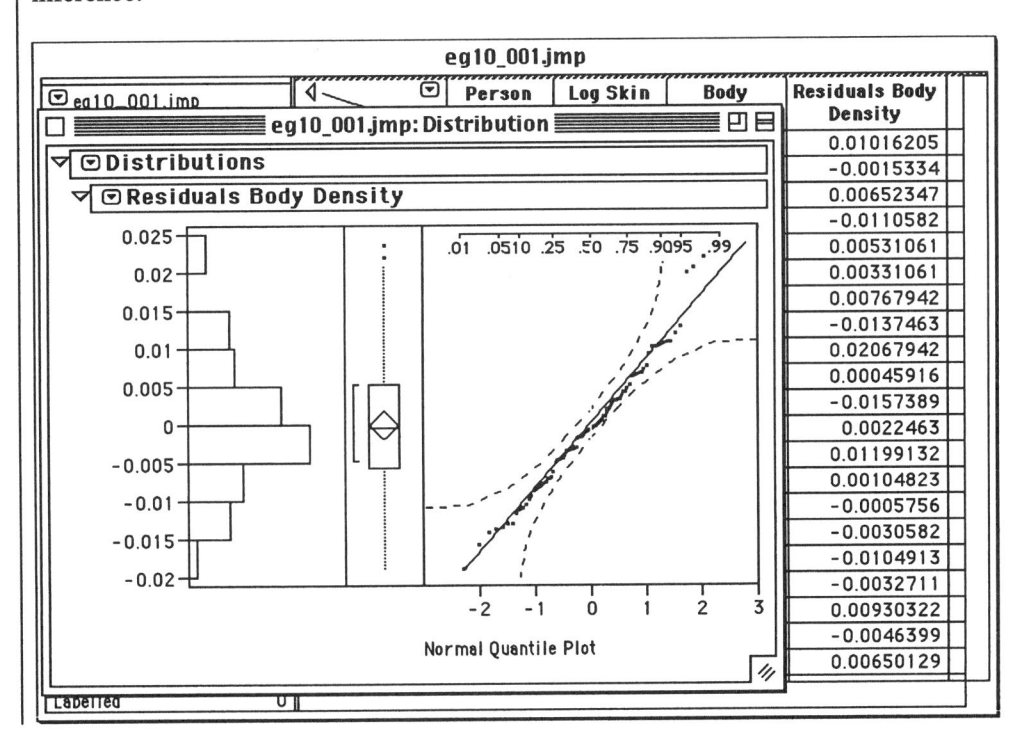

Plot the Residuals Against Other Variables

We should also plot the residuals against any other known variables; e.g., the individual (or case number) or the time or order in which the observations were taken. Since the residuals have been saved to the JMP data table, we can use all of the commands in the **Analyze** menu on the residuals.

6. Select **Analyze** ⇨ **Fit Y by X**.
   a. Select **Residuals Body Density** and press **Y, Columns**.
   b. Select **Person** and press **X, Columns** and **OK**.

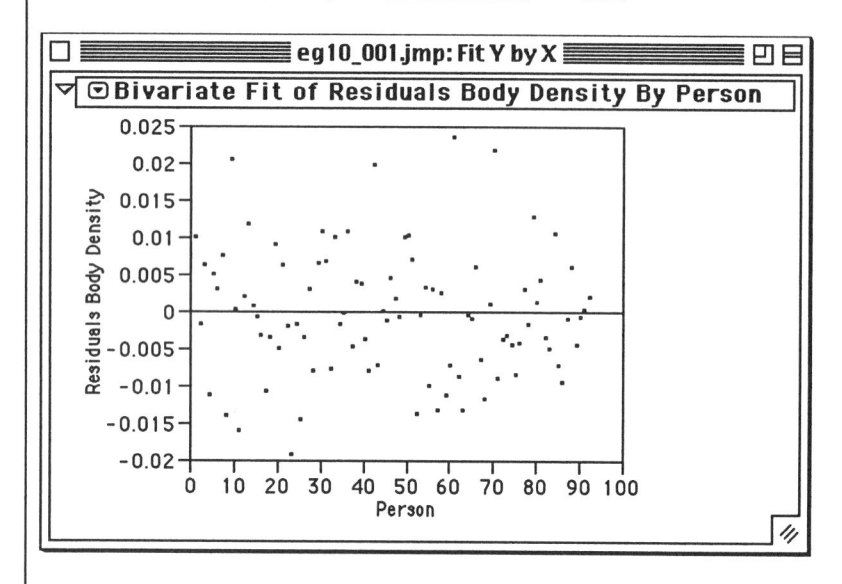

```
eg10_001.jmp: Fit Y by X
Bivariate Fit of Residuals Body Density By Person
```
[Scatter plot of Residuals Body Density vs Person, y-axis from -0.02 to 0.025, x-axis from 0 to 100]

No unusual patterns are evident. To add the reference line at 0, context-click on the **Residuals Body Density** axis and select **Axis settings** (see Section 0.6.2 of Chapter 0).

10.1.2 Inference for the Regression Slope and Intercept

Having determined that the assumptions of the model hold, we consider extending our conclusions about the relation of body density to skinfold thickness beyond the 92 males in the sample.

IPS Example 10.3 Body density and skinfold thickness (cont'd.)

Use the **Window** menu to bring the **eg10_001.jmp:Fit Y by X** window forward.

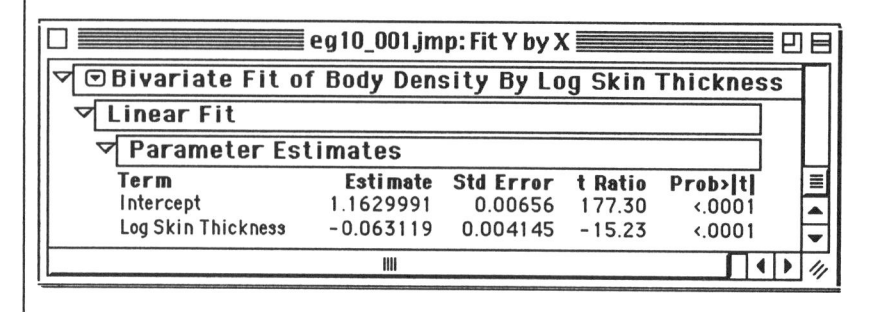

```
eg10_001.jmp: Fit Y by X
Bivariate Fit of Body Density By Log Skin Thickness
Linear Fit
Parameter Estimates
```

Term	Estimate	Std Error	t Ratio	Prob>|t|
Intercept	1.1629991	0.00656	177.30	<.0001
Log Skin Thickness	-0.063119	0.004145	-15.23	<.0001

In the **Parameter Estimates** section,

- The **Estimate** column contains the least-squares estimates for the unknown regression parameters. The estimate of the *slope* ß$_1$ (–0.0631) appears in the 'Log Skin thickness' row.

- The **Std Error** column contains the standard errors. In particular, SE$_{b1}$ = 0.004145.

- To test the hypothesis H$_o$: the true *slope* is zero (ß = 0) against H$_a$: ß$_1$ ≠ 0, JMP calculates the *t* statistic (labeled **t Ratio**) and the corresponding *P*-value (labeled **Prob>|t|**) for the two-sided alternative. To test the hypothesis,

    H$_o$: **Log Skin Thickness** is of no value in predicting **Body Density** (ß$_1$ = 0) versus
    H$_a$: **Log Skin Thickness** is linearly related to **Body Density** (ß$_1$ ≠ 0),

    the *t* statistic = –15.23 and the *P*-value < 0.0001. There is very strong evidence that **Log Skin Thickness** is related to **Body Density**.

Corresponding information is also given for the intercept.

JMP does not give a confidence interval estimate for the true slope ß$_1$ (or the intercept ß$_0$). However, you can easily calculate one using the information given above and the critical value *t** from Table C of IPS.

10.1.3 Inference about Prediction

JMP displays a graph of the *confidence intervals for mean responses* in the form of two curves around the least-squares regression line. To read the confidence interval for the mean of the response *y* when *x* has the value *x** from the graph, use the **Crosshair tool**. *Prediction intervals for individual responses* are handled similarly.

IPS Example 10.4 Predicting body density (cont'd.)

Find a 95% confidence interval for the mean body density of all men aged 20 to 29 whose logarithm of the sum of the skinfold thickness measures is 1.27.

The least-squares line is **Body Density**^ = 1.1629991 – 0.0631 **Log Skin Thickness**. There is strong evidence that skinfold thickness is useful in predicting body density (*t* = –15.23, *P*-value < .0001). Therefore, we estimate the mean body density of all men aged 20 to 29 whose **Log Skin Thickness** is 1.27 to be 1.1629991 – 0.063119 (1.27) = 1.0828. To obtain the 95% *confidence interval* for the mean body density of these men,

1. Press the red triangle below the plot next to **Linear Fit** and select <u>**Confid Curves Fit**</u> from the menu that opens.

The curves above and below the least-squares line give the 95% *confidence intervals for mean responses*.

[Figure: Bivariate Fit of Body Density By Log Skin Thickness — scatterplot with linear fit and confidence curves; x-axis Log Skin Thickness from .9 to 2.1, y-axis Body Density from 1.02 to 1.1]

2. Select **Tools** ⇨ (the Crosshair tool).

3. Place the crosshair on the lower curve directly above **Log Skin Thickness** = 1.27 to obtain the value of the lower confidence limit. Press and hold the mouse button. The values of the *x* and *y* axes where the crosshair intersects the plot appear automatically; in this case, 1.27 and 1.0800, respectively. You may need to drag the crosshair a bit.

4. Obtain the upper confidence limit by placing the crosshair on the upper curve directly above **Log Skin Thickness** = 1.27. A value of 1.08578 for **Body Density** is displayed.

Hence, the 95% *confidence interval* for the mean **Body Density** of all men aged 20 to 29 with a **Log Skin Thickness** of 1.27 is approximately (1.080, 1.086).

The 95% *prediction interval* for the **Body Density** of an individual with **Log Skin Thickness** of 1.27 is obtained in a similar way.

5. Select **Confid Curve Indiv** from the red triangle menu next to **Linear Fit**. The curves further from the least-squares line give the 95% *prediction intervals for individual responses*.

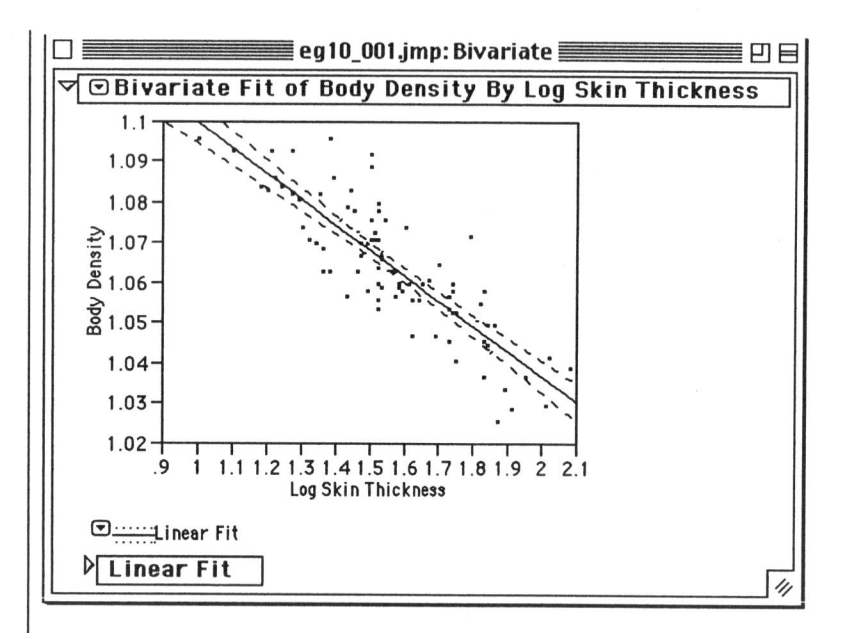

Simple Linear Regression

[Figure: Bivariate Fit of Body Density By Log Skin Thickness showing scatterplot with linear fit line and confidence/prediction bands. X-axis: Log Skin Thickness (0.9 to 2.1); Y-axis: Body Density (1.02 to 1.1)]

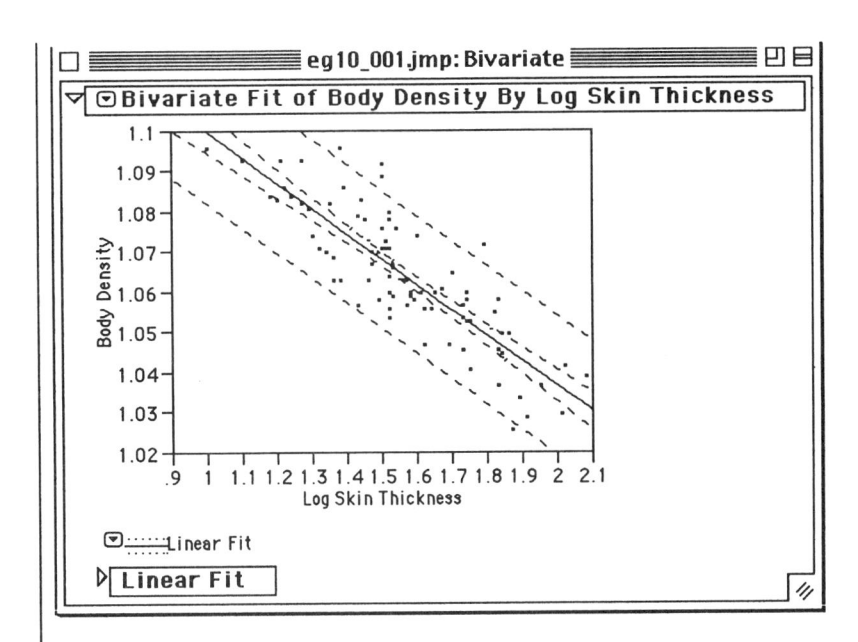

6. Use the crosshair tool to estimate the upper and lower prediction limits when **Log Skin Thickness** is 1.27.

The 95% *prediction interval* for the **Body Density** of an individual with a **Log Skin Thickness** measure of 1.27 (1.066, 1.100) is much wider than the *confidence interval* for the mean.

Remarks

- To obtain prediction intervals and confidence intervals for levels of confidence other than 95%, use the **Set Alpha Level** command on the **Linear Fit** red triangle menu, and choose the value $1 - C$ where C is the level desired.

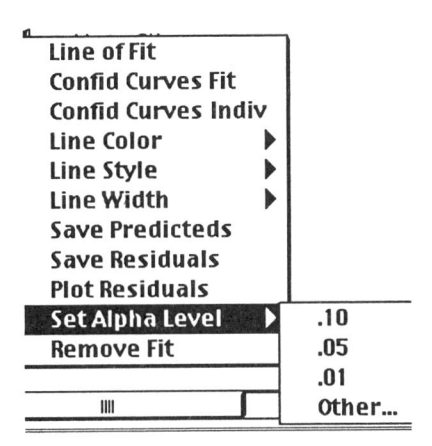

[Menu image showing: Line of Fit, Confid Curves Fit, Confid Curves Indiv, Line Color, Line Style, Line Width, Save Predicteds, Save Residuals, Plot Residuals, Set Alpha Level (highlighted with submenu: .10, .05, .01, Other...), Remove Fit]

- Predicted values for the sample data can be saved to the JMP data table. The **Save Predicteds** command on the **Linear Fit** red triangle menu creates a new column that contains the sample predicted values. To obtain predicted values for other *x* values, simply add rows, containing only the *x* values, to the table.

10.2 More Detail about Simple Linear Regression

10.2.1 The ANOVA F Test

JMP provides the analysis of variance table and F statistic as part of the Linear Fit report.

IPS Example 10.7 Body density and skinfold thickness (cont'd.)

Use the **Window** menu to bring the **eg10_001.jmp:Fit Y by X** window forward. (If it is closed, repeat step 2 of IPS Example 10.1.)

Bivariate Fit of Body Density By Log Skin Thickness
Linear Fit
Analysis of Variance

Source	DF	Sum of Squares	Mean Square	F Ratio
Model	1	0.01690856	0.016909	231.8942
Error	90	0.00656235	0.000073	Prob > F
C. Total	91	0.02347090		<.0001

The report matches the ANOVA table format given in the IPS textbook and also provides the *P*-value (labeled **Prob > F**).

10.2.2 Inference for Correlation

To compute the sample correlation and to test the hypothesis H_0: $\rho = 0$ against the alternative hypothesis H_a: $\rho \neq 0$, we use the **Density Ellipse** command on the **Fit Y by X** platform in JMP INTRO. (See Section 2.2 in Chapter 2.)

IPS Example 10.15 Body density and skinfold thickness (cont'd.)

Use the **Window** menu to bring the **eg10_001.jmp:Fit Y by X** window forward. (If it is closed, repeat step 2 of IPS Example 10.1.)

1. Press the red triangle next to **Bivariate Fit of Body Density By Log Skin Thickness** and select **Density Ellipse** ⇨ **0.95**. (It doesn't matter which number you select.)

2. Open the **Correlation** report by clicking the pale blue diamond-shaped disclosure button next to the title **Correlation**.

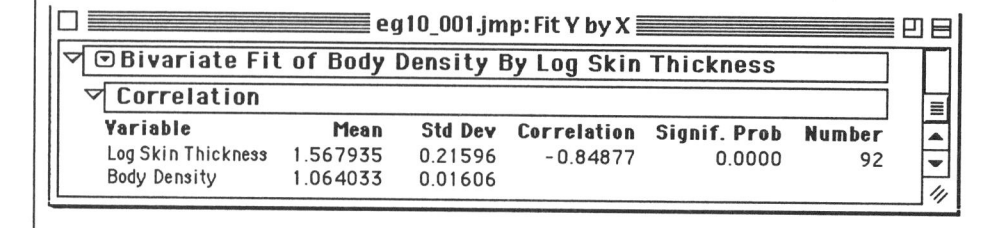

Bivariate Fit of Body Density By Log Skin Thickness
Correlation

Variable	Mean	Std Dev	Correlation	Signif. Prob	Number
Log Skin Thickness	1.567935	0.21596	-0.84877	0.0000	92
Body Density	1.064033	0.01606			

Notice that the sample correlation between body density and log of skinfolds is $r = -0.84877$. The number 0.000 directly below **Signif. Prob** is the *P*-value for testing the two-side alternative hypothesis that the population correlation is not 0. Since the P-value is so small, the data provide strong evidence that there is a negative correlation between body density and log of skinfolds.

Remark

In JMP IN and the professional version of JMP, the **Multivariate** analysis platform can also be used to test for a zero population correlation.

10.3 Summary

All graphs and statistical computations to perform inference for regression use the **Fit Y by X** platform. The residuals can be examined using the **Distribution** platform.

Activity	Command
Inference about the model	**Fit Y by X** ⇨ **Fit Line**
Inference about prediction	**Fit Y by X** ⇨ **Fit Line** ⇨ ...
To estimate an individual response	**Confid Curves Indiv**
To estimate the mean response	**Confid Curves Fit**
Checking assumptions	**Fit Y by X** ⇨ **Fit Line** ⇨ **Plot Residuals**
	Fit Y by X ⇨ **Fit Line** ⇨ **Save Residuals**
Inference for correlation	**Fit Y by X** ⇨ **Density Ellipse**

10.4 Exercises

You should use JMP on any problem in the IPS text for which the data are provided. Here are some suggestions:

1. Beer and blood alcohol levels. Exercises 10.4, 10.6, and 10.28.

2. Growth of DRAM capacity. Exercise 10.5.

3. Bank wages and length of service. Exercises 10.9 and 10.10.

4. Predicting oxygen uptake. Exercises 10.20 and 10.32.

5. Speed and gas mileage. Exercise 10.47.

6. Do heavy people burn more energy? Exercises 10.48 and 10.49. The **By** option on the **Fit Y by X** platform may be of help in fitting models for men and women separately.

7. Fish sizes. Exercise 10.24.

8. Fish weights. Exercise 10.25.

Chapter 11

Multiple Regression

Chapter 10 discussed inference for a linear relationship between two quantitative variables—a response variable and a *single explanatory* variable. Chapter 11 considers the case when there are *multiple explanatory* variables.

11.1 Inference for Multiple Regression

In this section of the IPS textbook, the *multiple linear regression* model and methods for estimating the unknown population regression parameters are discussed. Methods of inference for the regression coefficients and the analysis of variance F test are also presented. In the next section, we use JMP to fit and analyze the multiple regression model.

11.2 A Case Study

When there are several explanatory variables, often the goal is to determine what subset of the explanatory variables best predicts the response variable. Thus, in addition to simply fitting and testing a multiple regression model, we illustrate the process of model building using JMP. First, we will examine the distribution of each variable separately. Next, we examine the relationships between pairs of variables. Then, we fit and test the full multiple regression model and various subset models. We illustrate the process with a case study.

IPS Example 11.1 Predicting academic success in college

The Computer Science Department at a university wished to understand why a large percentage of its first-year students failed to graduate as computer science majors. The purpose of the study was to attempt to predict success in the early university years as measured by the cumulative grade point average (GPA) after three semesters. Among the explanatory variables, measured at the time the students enrolled in the university, were the average high school grades in mathematics (HSM), science (HSS), and English (HSE). The text file **CSDATA.TXT** in the Appendix folder on the IPS CD-ROM contains the data.

1. Import the text file **CSDATA.TXT** on the IPS CD-ROM and name the variables **Obs**, **GPA**, **HSM**, **HSS**, **HSE**, **SATM**, **SATV**, and **Sex.**

There should be 224 individuals and 8 variables.

2. Save the JMP data table as **csdata.jmp**.

Distributions of Each Variable Separately

Let's use the **Distribution** platform in JMP to look at the distributions of the quantitative variables.

3. Select **Analyze** ⇨ **Distribution**.

4. Select **GPA**, **SATM**, **SATV**, **HSM**, **HSS**, and **HSE**; press **Y, Columns** and **OK**.

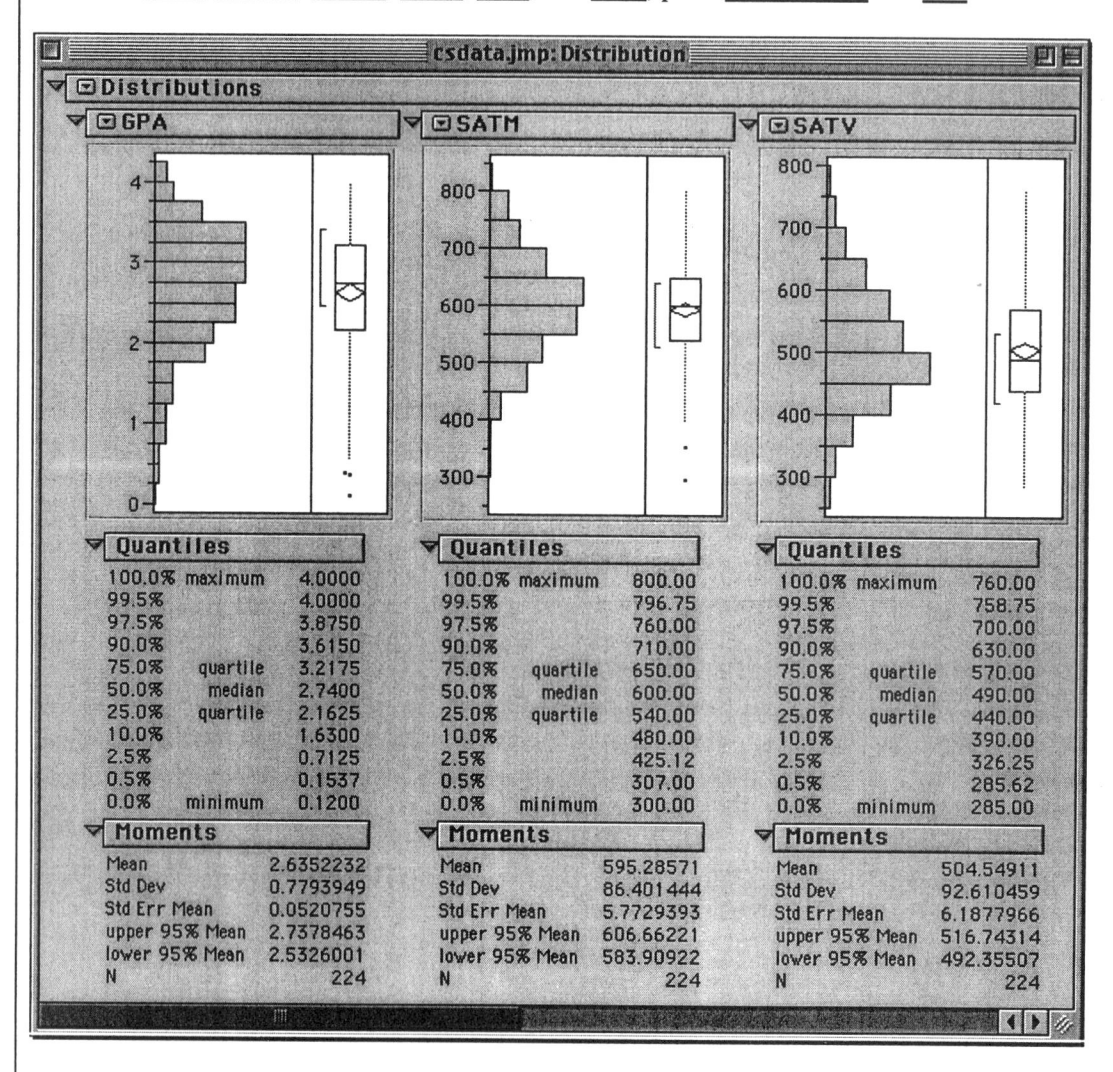

	GPA	SATM	SATV
100.0% maximum	4.0000	800.00	760.00
99.5%	4.0000	796.75	758.75
97.5%	3.8750	760.00	700.00
90.0%	3.6150	710.00	630.00
75.0% quartile	3.2175	650.00	570.00
50.0% median	2.7400	600.00	490.00
25.0% quartile	2.1625	540.00	440.00
10.0%	1.6300	480.00	390.00
2.5%	0.7125	425.12	326.25
0.5%	0.1537	307.00	285.62
0.0% minimum	0.1200	300.00	285.00

Moments	GPA	SATM	SATV
Mean	2.6352232	595.28571	504.54911
Std Dev	0.7793949	86.401444	92.610459
Std Err Mean	0.0520755	5.7729393	6.1877966
upper 95% Mean	2.7378463	606.66221	516.74314
lower 95% Mean	2.5326001	583.90922	492.35507
N	224	224	224

The response variable **GPA** is skewed to smaller numbers with a few students who have probably given up. The five-number summary is 0.12, 2.16, 2.74, 3.21, 4.00. A lack of normality of the variable **GPA** is not of concern. Only the residuals of a fitted model must be normally distributed. Examine the distributions of the **SATM** and **SATV** scores to learn something about them alone before attempting to use them in a complex model.

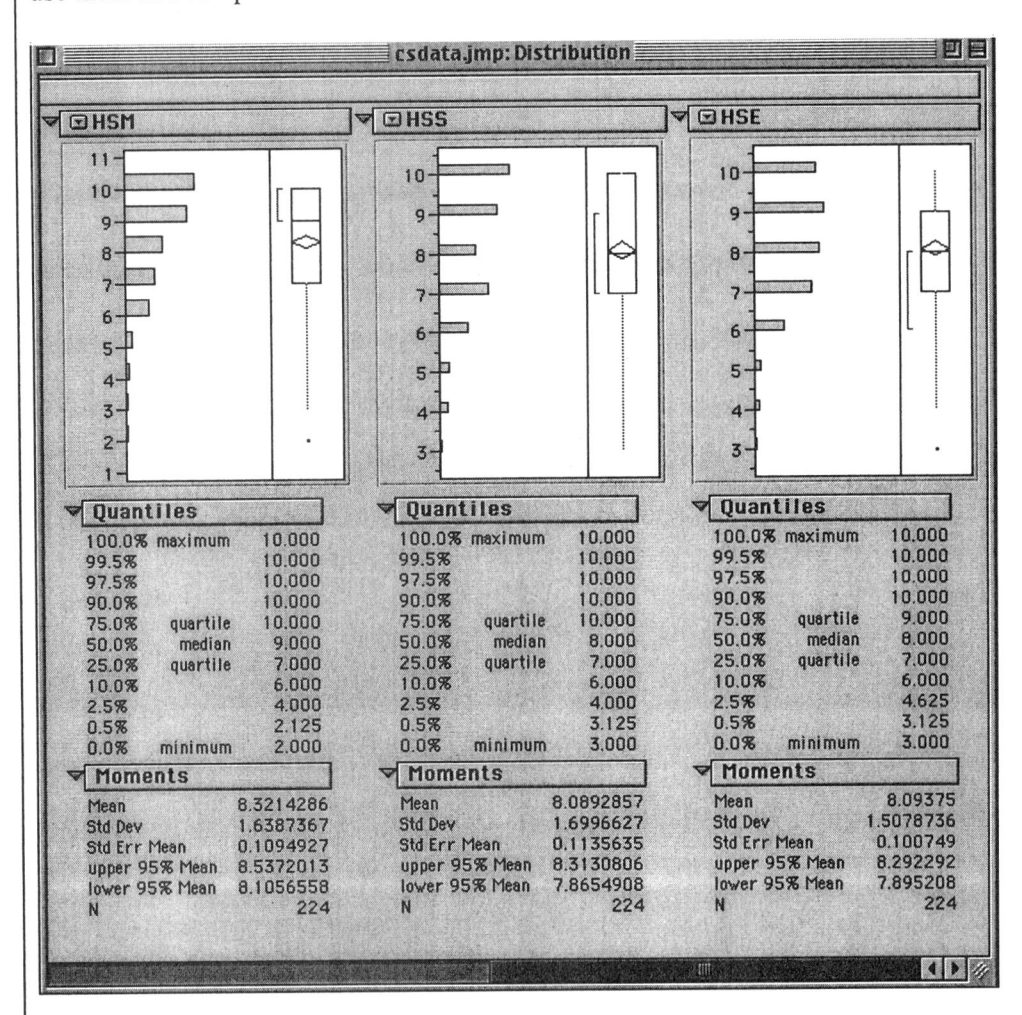

[JMP Distribution output window for csdata.jmp showing histograms, box plots, Quantiles, and Moments for HSM, HSS, and HSE:

HSM — Quantiles: 100.0% maximum 10.000; 99.5% 10.000; 97.5% 10.000; 90.0% 10.000; 75.0% quartile 10.000; 50.0% median 9.000; 25.0% quartile 7.000; 10.0% 6.000; 2.5% 4.000; 0.5% 2.125; 0.0% minimum 2.000. Moments: Mean 8.3214286; Std Dev 1.6387367; Std Err Mean 0.1094927; upper 95% Mean 8.5372013; lower 95% Mean 8.1056558; N 224.

HSS — Quantiles: 100.0% maximum 10.000; 99.5% 10.000; 97.5% 10.000; 90.0% 10.000; 75.0% quartile 10.000; 50.0% median 8.000; 25.0% quartile 7.000; 10.0% 6.000; 2.5% 4.000; 0.5% 3.125; 0.0% minimum 3.000. Moments: Mean 8.0892857; Std Dev 1.6996627; Std Err Mean 0.1135635; upper 95% Mean 8.3130806; lower 95% Mean 7.8654908; N 224.

HSE — Quantiles: 100.0% maximum 10.000; 99.5% 10.000; 97.5% 10.000; 90.0% 10.000; 75.0% quartile 9.000; 50.0% median 8.000; 25.0% quartile 7.000; 10.0% 6.000; 2.5% 4.625; 0.5% 3.125; 0.0% minimum 3.000. Moments: Mean 8.09375; Std Dev 1.5078736; Std Err Mean 0.100749; upper 95% Mean 8.292292; lower 95% Mean 7.895208; N 224.]

The high school grade variables **HSM**, **HSS**, and **HSE** are discrete variables with ordered values. If we temporarily change their modeling type to "Ordinal," JMP will display a mosaic plot and frequency table for each.

5. Select the columns **HSM**, **HSS**, and **HSE** of **csdata.jmp** in the JMP data table window.
   a. Select **Cols** ⇨ **Col Info**.
   b. Select **Ordinal** in each of the `Modeling Type` fields and press **OK**.

6. Select **Analyze** ⇨ **Distribution**.

7. Select **HSM**, **HSS**, and **HSE**; press **Y, Columns** and **OK**.

[JMP distribution output for HSM, HSS, and HSE]

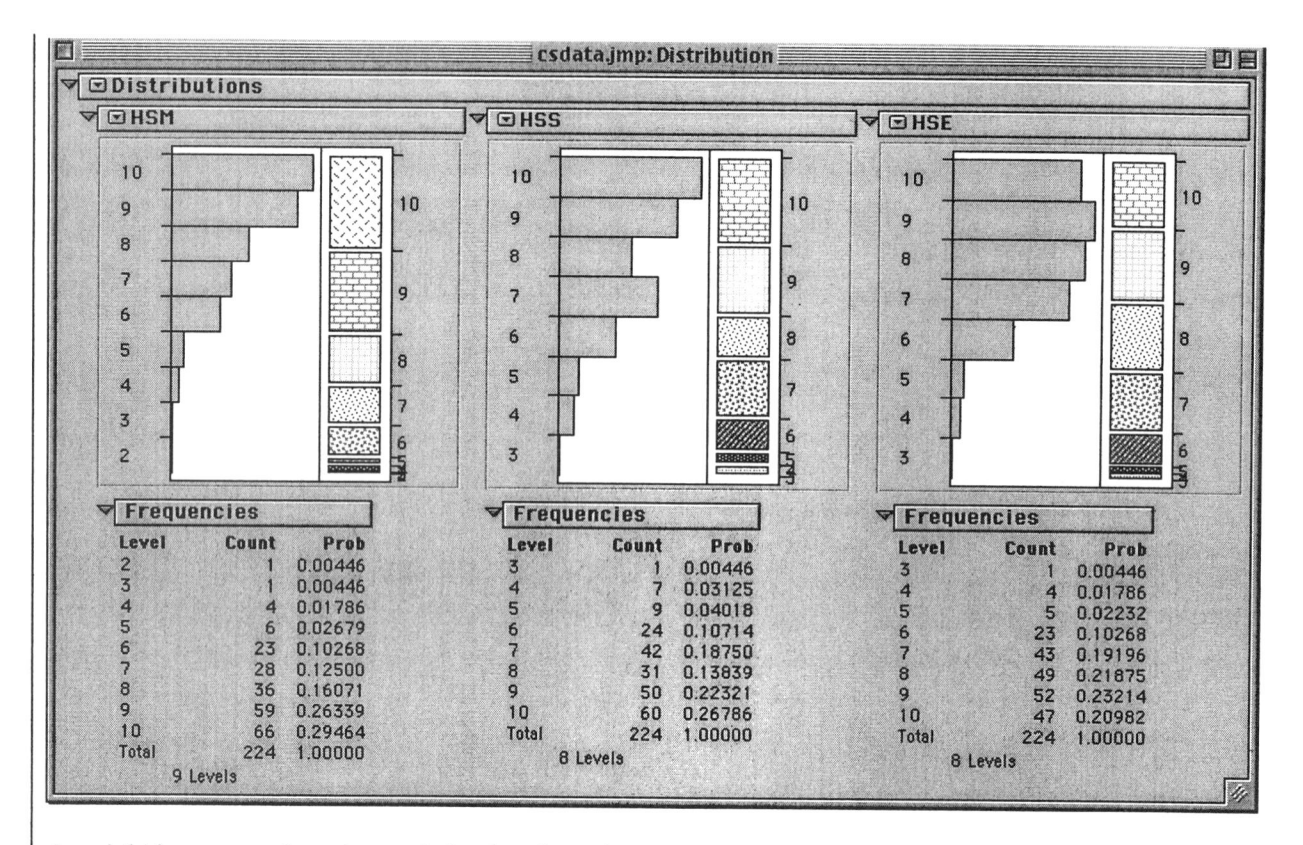

As might be expected, students admitted to the university did well in high school and the high school grade variables are skewed toward the lower grades.

Relationships Between Pairs of Variables

Let's use the **Fit Y by X** platform to look for two-variable relationships. First, be sure to return the modeling type of **HSM**, **HSS**, and **HSE** to "Continuous."

1. Repeat step 5 from the previous page but select **Continuous** instead of Ordinal.

2. Select **Analyze** ⇨ **Fit Y by X**.

3. Select **GPA**, **SATM**, **SATV**, **HSM**, **HSS**, and **HSE**; press **Y, Response**, **X, Factor**, and **OK**.

We use a shortcut to obtain the pairwise correlations.
   a. Hold down the Ctrl key (Command key on the Macintosh).
   b. Click on the red triangle on the title of any scatterplot and select **Density Ellipse** ⇨ **.95**.
   c. Simultaneously press the Ctrl key and the disclosure button for a **Correlation** report and select **Open All Like This**.

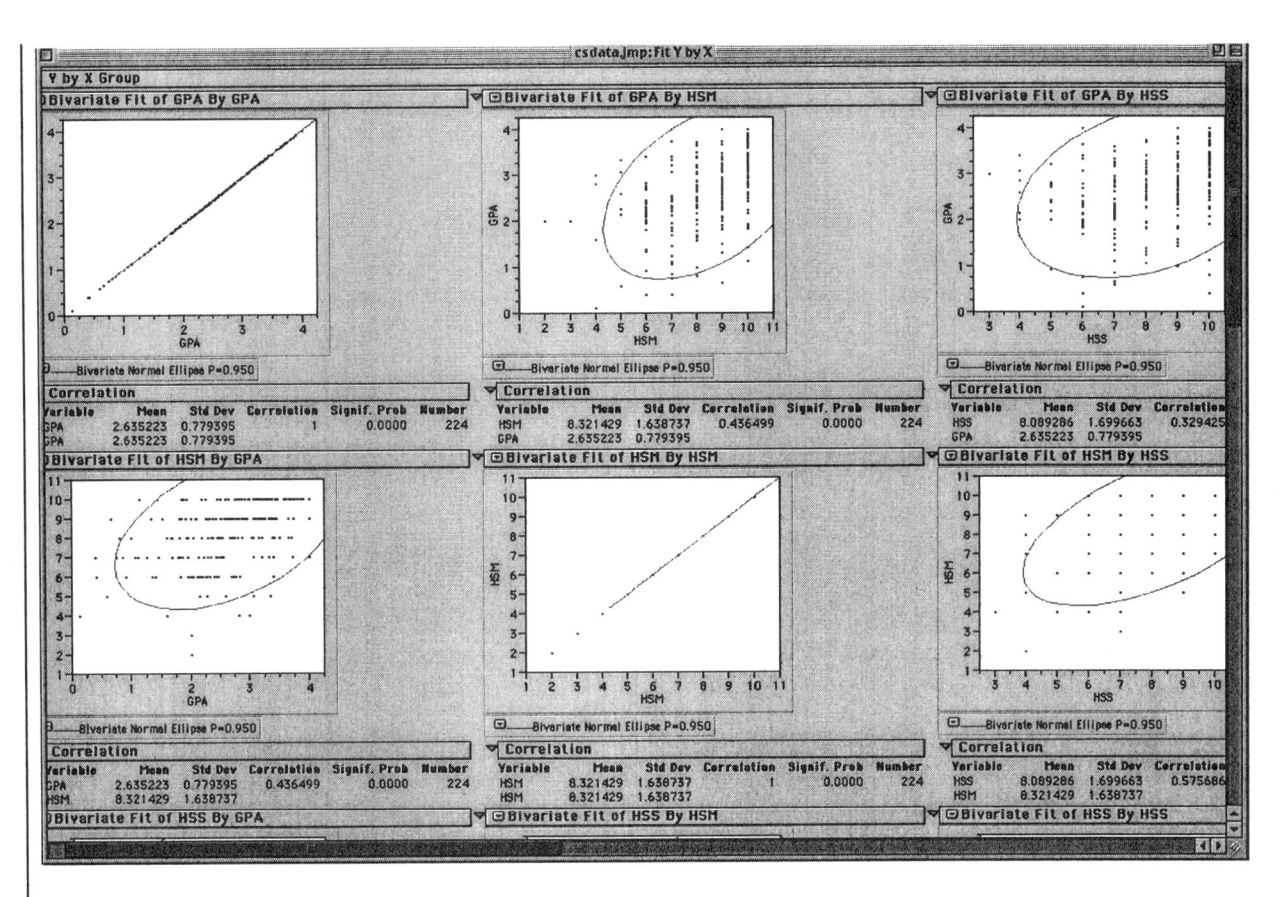

Examine the pairwise correlations and scatterplots. The high school grades are more correlated with GPA than the SAT scores. The high school grades are correlated with one another, as are the two SAT scores.

Remark

In JMP IN and the professional version of JMP, the **Correlation** command in the **Multivariate** analysis platform provides a succinct summary of the pairwise relations among variables.

1. Select **Analyze** ⇨ **Multivariate.**

2. Select **GPA**, **SATM**, **SATV**, **HSM**, **HSS**, and **HSE**; press **Y, Columns** and **OK**.

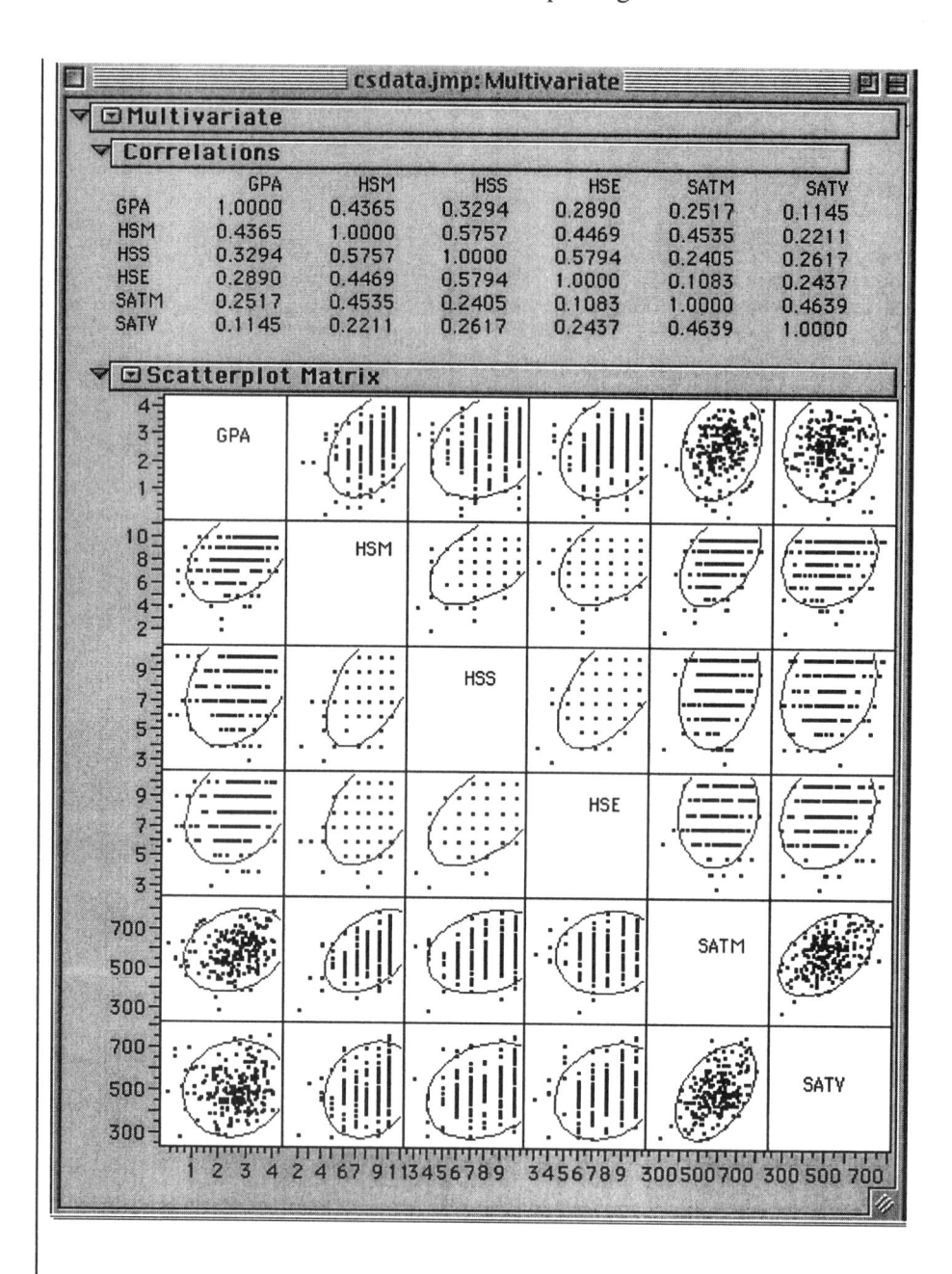

```
                csdata.jmp: Multivariate
▽ Multivariate
  ▽ Correlations
            GPA     HSM     HSS     HSE     SATM    SATV
   GPA    1.0000  0.4365  0.3294  0.2890  0.2517  0.1145
   HSM    0.4365  1.0000  0.5757  0.4469  0.4535  0.2211
   HSS    0.3294  0.5757  1.0000  0.5794  0.2405  0.2617
   HSE    0.2890  0.4469  0.5794  1.0000  0.1083  0.2437
   SATM   0.2517  0.4535  0.2405  0.1083  1.0000  0.4639
   SATV   0.1145  0.2211  0.2617  0.2437  0.4639  1.0000
```

Regression on High School Grades

The first multiple regression model to be studied uses only the three high school grades to predict GPA. We use a different JMP analysis platform (**Fit Model**) to fit the multiple regression model.

1. Select **Analyze** ⇨ **Fit Model**.
   a. Select **GPA** from the list of columns and press **Y**.
   b. Select **HSM**, **HSS**, and **HSE**; press **Add**.
   c. Press **Run Model**.

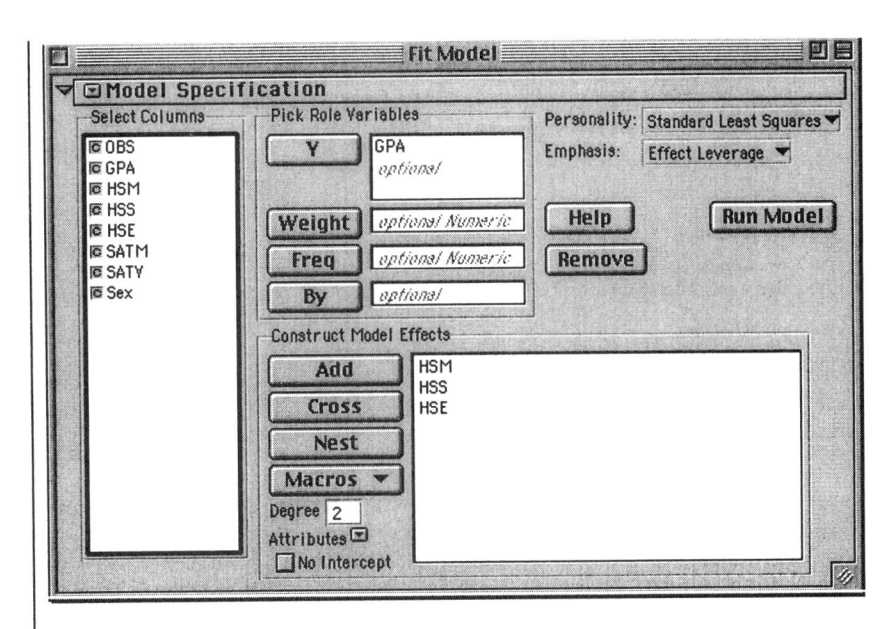

The **Fit Model** report window opens. Examine the **Whole Model** report. The degrees of freedom in the **Analysis of Variance** report are as expected with 220 degrees of freedom for error. The ANOVA F statistic is 18.8606 with a P-value less than 0.0001. We can conclude that at least one of the population regression coefficients for the high school grades is different from zero.

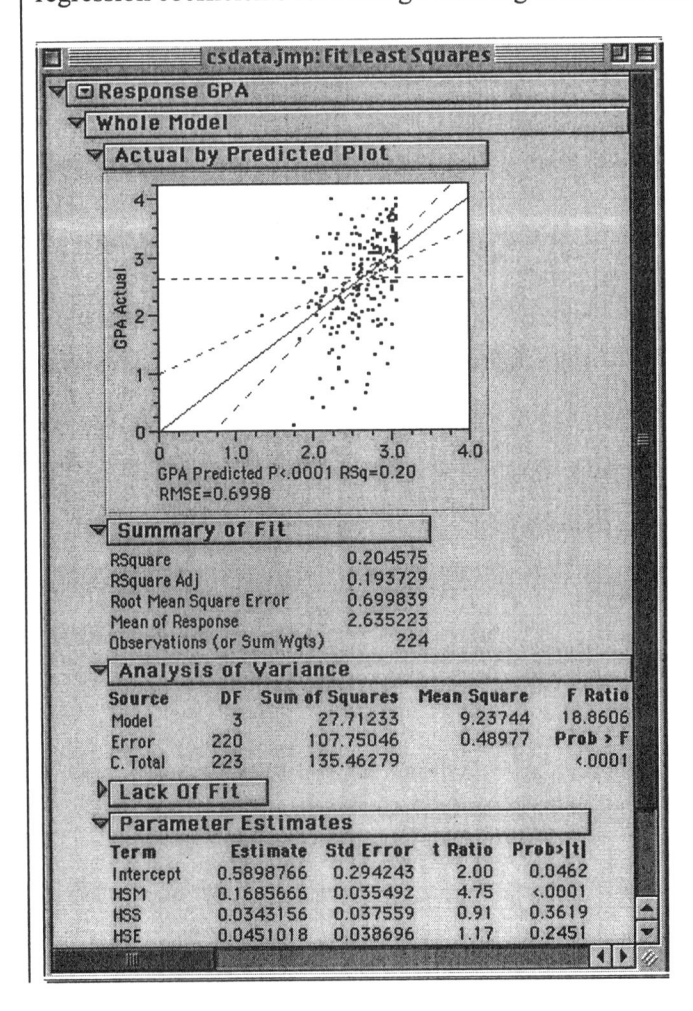

Investigate the **Summary of Fit** report. The root mean square error (or s, the estimate of the parameter σ) is 0.699839. The squared multiple correlation R^2 (labeled **RSquare**) equals 0.204575. Although significant, the model explains only about 20% of the variability in the **GPA** scores.

From the **Parameter Estimates** report, we obtain the fitted regression equation

$$\widehat{\text{GPA}} = 0.5899766 + 0.1685666\ \text{HSM} + 0.0343156\ \text{HSS} + 0.0451018\ \text{HSE}$$

The *t statistics* for testing the regression coefficients and their associated *P*-values appear in the last two columns labeled **t Ratio** and **Prob>|t|**, respectively. Only the coefficient of **HSM** (high school math grade) is significantly different from zero.

These results are supported by the associated graphics. The plot shown at the top of the **Whole Model** report shows actual GPAs plotted versus GPAs predicted by the model. A diagonal 45° reference line and 95% confidence intervals for the model fit are also plotted. Because the confidence interval does not contain the blue line that represents the no effect model, you conclude that the whole model is statistically significant. This confirms the ANOVA *F* test result.

JMP also produces *leverage plots*, or *added variable plots*, for each explanatory variable. They confirm the results of the *t* statistics. The plots enable you to see the residual of the full model and the residual of the model without the variable of interest. A graphical test of whether a variable is important is obtained by comparing the confidence curves about the red line to the horizontal blue line just as in the Whole Model test. Only the confidence interval in the leverage plot for **HSM** does not contain the blue line. Therefore, only the coefficient of **HSM** is significantly different from zero.

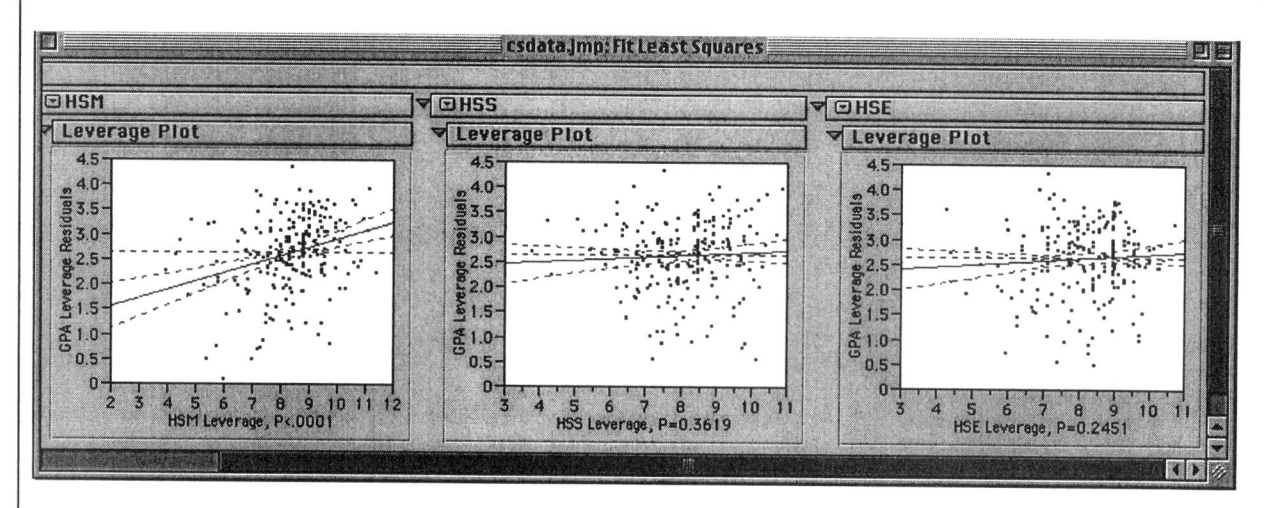

Residuals

A plot of the residuals versus the predicted values can be found at the bottom of the **Whole Model** report. It appears to be random noise.

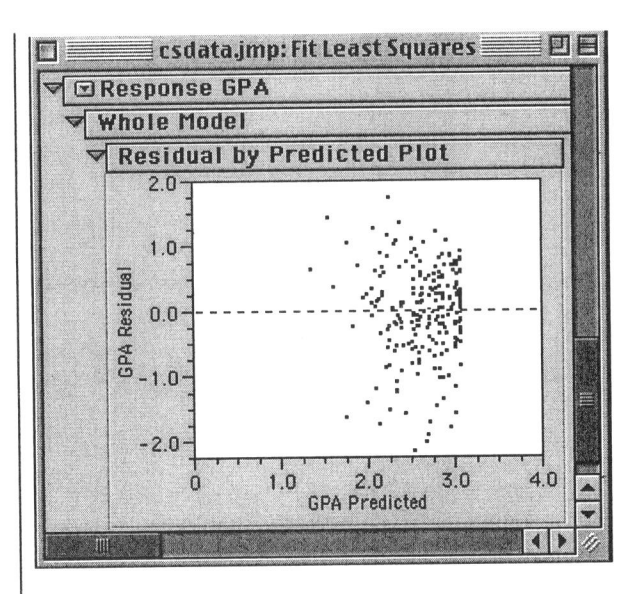

[Screenshot: csdata.jmp: Fit Least Squares — Residual by Predicted Plot showing GPA Residual vs GPA Predicted]

To obtain a normal quantile plot of the residuals, we save the residuals to the data table and then use the **Normal Quantile Plot** command in the **Distribution** platform.

1. Click on the red triangle next to **Response GPA**.

2. Select **Save Columns** ⇨ **Residuals** from the menu that opens.

3. Select **Analyze** ⇨ **Distribution**.
   a. Select **Residual GPA**, and press **Y, Columns** and **OK**.
   b. Select **Normal Quantile Plot** from the red triangle menu of the **Residual GPA** report.

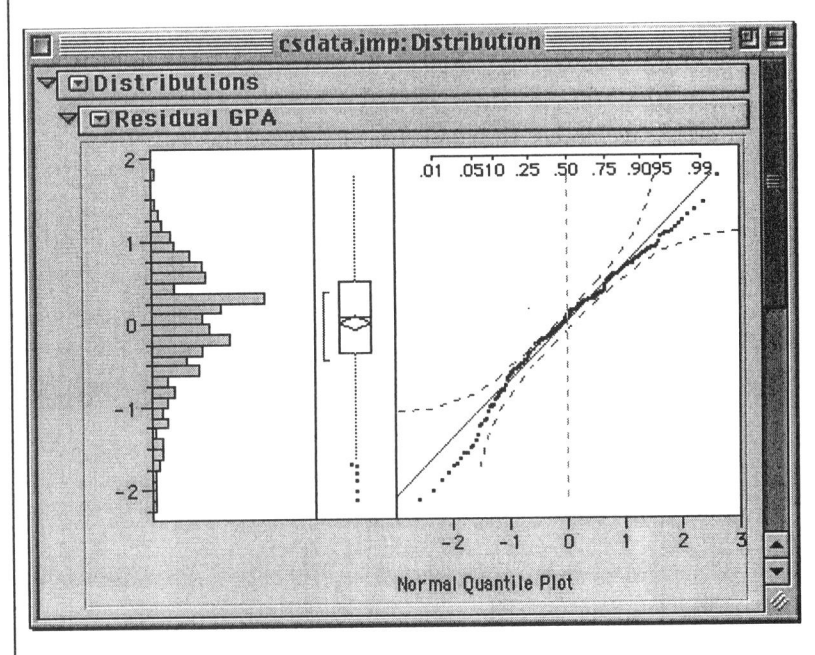

[Screenshot: csdata.jmp: Distribution — Residual GPA with histogram, boxplot, and Normal Quantile Plot]

While there appears to be some modest departures from normality in the center of the distribution, they are not likely to effect the *P*-values in view of the large sample size.

11.3 Summary

All graphs and statistical computations for multiple regression models use the **Fit Model** platform. The residuals can be examined using the **Distribution** platform.

Activity	Command
Inference about the model	**Analyze ⇨ Fit Model**
Saving residuals	**Fit Model ⇨ Save Columns ⇨ Residuals**

11.4 Exercises

You should use JMP on any problem in the IPS text for which the data are provided. Here are some suggestions:

1. Predicting GPA. CSDATA. HSS was not significant for predicting GPA when both HSM and HSE were in the model. Rerun the regression using only HSM and HSE as explanatory variables. Compare the resulting model with the one with all high school grades.

2. Internet stock brokerages. Exercises 11.8 through 11.12.

3. Using SAT scores to predict GPA. Exercises 11.18 and 11.19.

4. Gender and predicting GPA. Exercises 11.30 and 11.31. The **By** option on the **Fit Model** window will be helpful.

5. Tasting cheddar cheese. Exercises 11.35 through 11.43.

6. Predicting corn yield. Exercises 11.44 through 11.50.

Chapter 12

One-Way Analysis of Variance

Chapter 7 provided tools for comparing the means of two groups or treatments. In this chapter, we compare any number of means using techniques that generalize the tools of Chapter 7. The same JMP commands for comparing two means, **Fit Y by X** ⇨ **Means/Anova/t-Test**, will allow us to compare more than two means.

12.1 Inference for Comparing Two or More Means

The JMP data table layout is the same as in Section 7.2 in Chapter 7. Each row contains only one experimental unit or individual. There is a column to identify the groups or treatments and a column for the response variable. The column that identifies the groups or treatments should be thought of as an explanatory variable, or factor. Then, comparing groups is equivalent to evaluating the relationship between the two variables and determining whether the response variable depends on the explanatory variable. Thus, we use the **Fit Y by X** platform in JMP.

IPS Example 12.6 Comparing reading groups

A study of reading comprehension in children compared three methods of instruction. Several pretest variables were measured before any instruction was given. One purpose of the pretest was to see if the three groups of children were similar in their comprehension skills. Sixty-six children were randomly assigned to three methods of instruction: basal, DRTA, and strategies. The text file **ta12_001.txt** on the IPS CD-ROM contains the data from this study.

1. Import the data from the file **ta12_001.txt** on the IPS CD-ROM and name the three columns **Subject, Group**, and **Pretest Score**.

2. Save the data table as **ta12_001.jmp**.

Let μ_1, μ_2, and μ_3 denote the population mean pretest score for all children who will receive the basal, DRTA, and strategies methods, respectively. Then, we wish to test the hypothesis

$H_o: \mu_1 = \mu_2 = \mu_3$ versus
H_a: not all of the μ_i are equal; i.e., at least two of the μ_i differ.

Inspect the Data

We first look at the variables separately.

3. Select **Analyze** ⇨ **Distribution**; then select **Group** and **Pretest Score** and press **Y, Columns**.

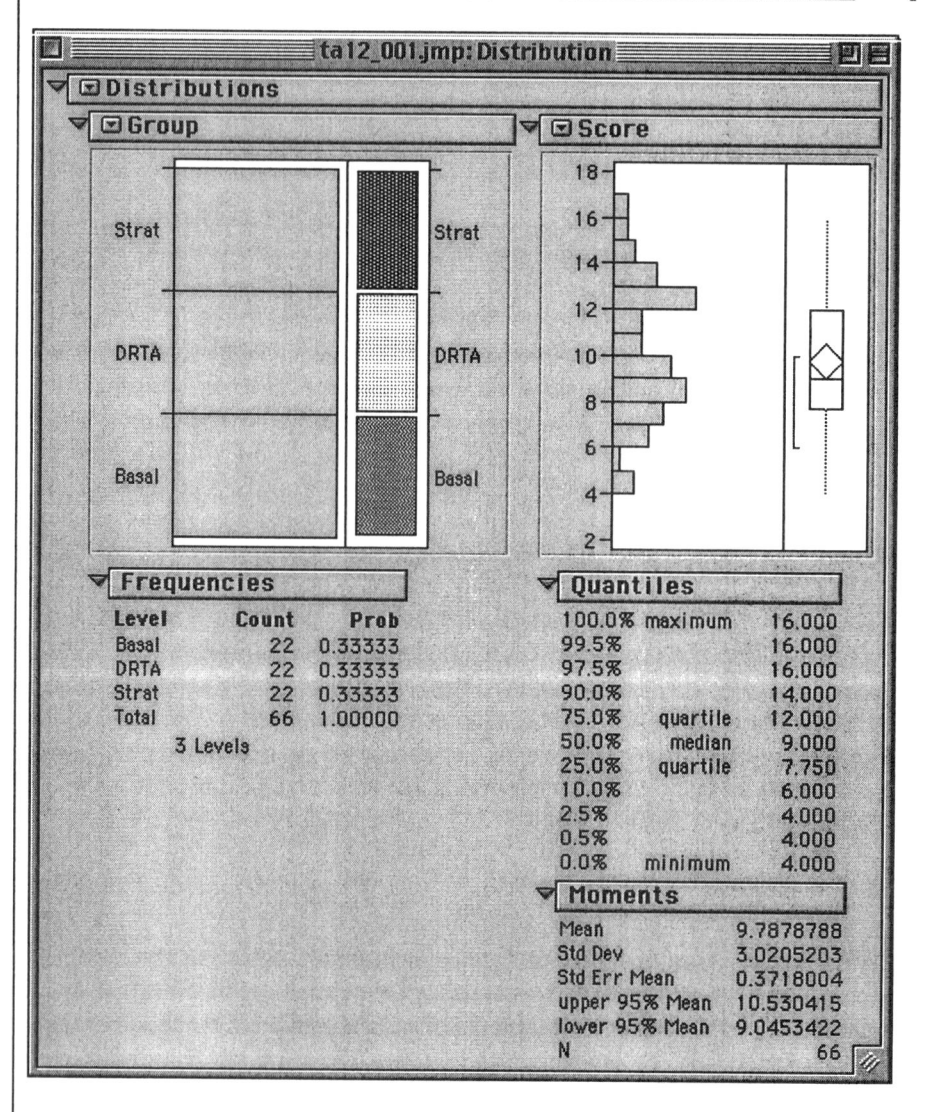

Each method of instruction will be given to 22 children. The mean pretest reading score is 9.79 with a standard deviation of 3.02. We now look at the variables together.

4. Select **Analyze** ⇨ **Fit Y by X**.
   a. Select **Pretest Score** and press **Y, Response**.
   b. Select **Group** and press **X, Factor** and **OK**.

The side-by-side point plots provide some insight into the effect of group on pretest reading score. However, we gain more by looking at side-by-side boxplots and side-by-side means diamonds.

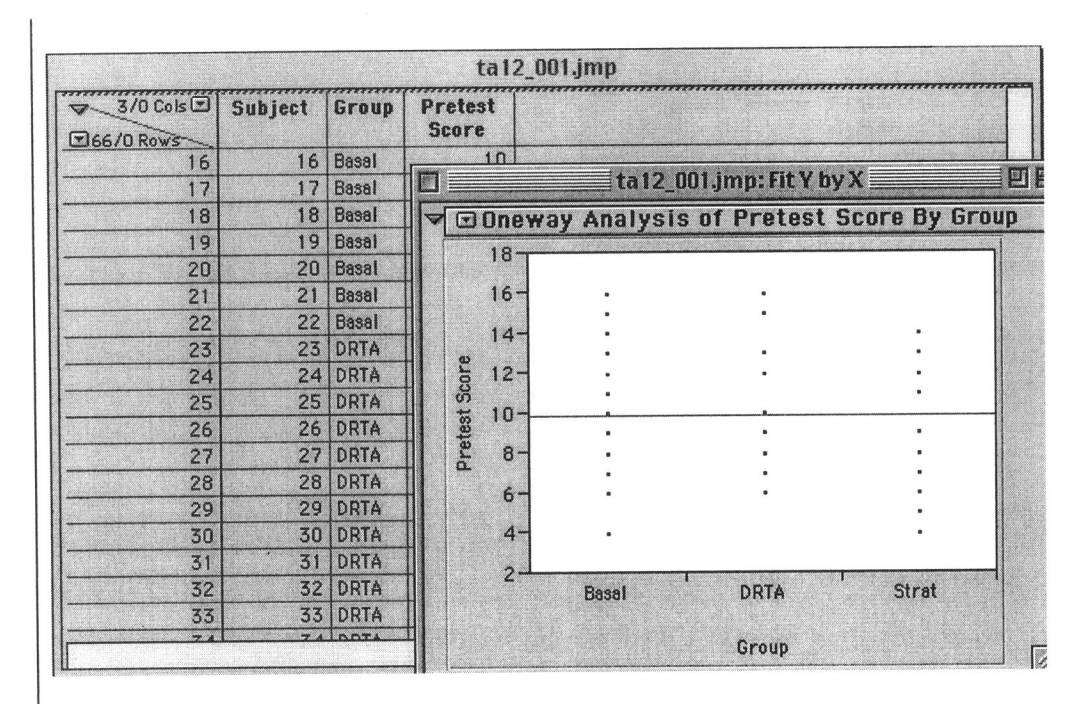

5. Select **Display Options** ⇨ **Box Plots** from the red triangle menu to display boxplots.

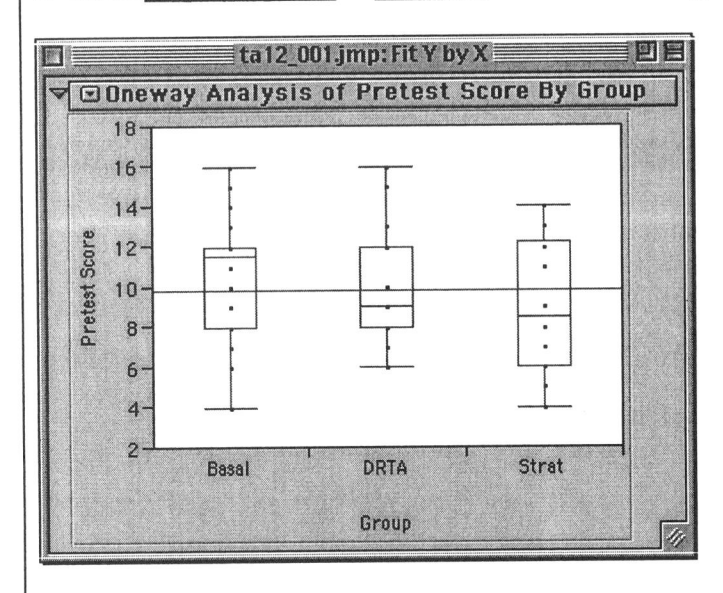

The average scores of the three groups do not differ by much.

Check Assumptions

In order to extend our results to the population means, certain assumptions must be satisfied. The textbook gives a useful rule of thumb for investigating departures from the assumption of equal variances. However, JMP allows us test that assumption. We used the command **UnEqual Variances** earlier in Section 7.3 in Chapter 7. We illustrate it again.

6. Click on the red triangle and select **UnEqual Variances**.

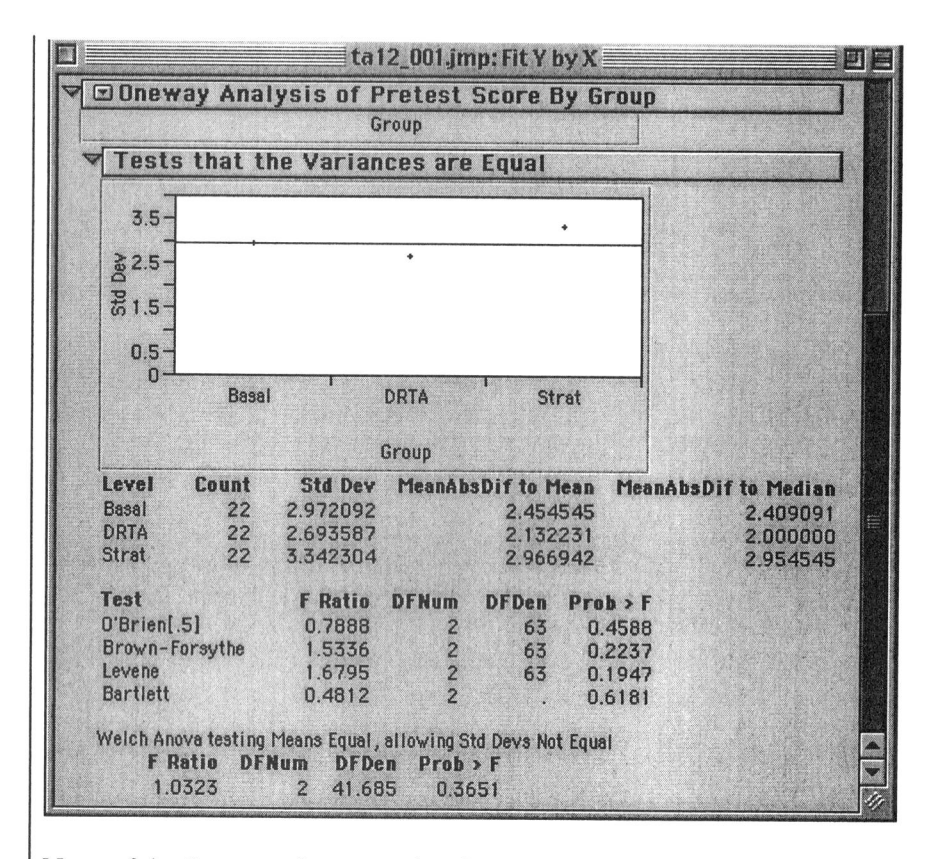

```
┌─────────────────────────────────────────────────────┐
│              ta12_001.jmp: Fit Y by X               │
│ ▽ Oneway Analysis of Pretest Score By Group         │
│                        Group                        │
│   ▽ Tests that the Variances are Equal              │
│                                                     │
│      3.5                              ·             │
│   S  3                                              │
│   t 2.5              ·                              │
│   d  2                                              │
│   D 1.5                                             │
│   e  1                                              │
│   v 0.5                                             │
│      0                                              │
│         Basal        DRTA         Strat             │
│                        Group                        │
│                                                     │
│   Level    Count   Std Dev   MeanAbsDif to Mean   MeanAbsDif to Median │
│   Basal    22      2.972092        2.454545            2.409091        │
│   DRTA     22      2.693587        2.132231            2.000000        │
│   Strat    22      3.342304        2.966942            2.954545        │
│                                                     │
│   Test            F Ratio   DFNum   DFDen   Prob > F│
│   O'Brien[.5]     0.7888      2       63    0.4588  │
│   Brown-Forsythe  1.5336      2       63    0.2237  │
│   Levene          1.6795      2       63    0.1947  │
│   Bartlett        0.4812      2        .    0.6181  │
│                                                     │
│   Welch Anova testing Means Equal, allowing Std Devs Not Equal │
│     F Ratio   DFNum    DFDen    Prob > F            │
│      1.0323     2      41.685    0.3651             │
└─────────────────────────────────────────────────────┘
```

None of the four tests for unequal variances is significant at the 0.10 level. JMP also provides the **Normal Quantile Plot** command to check for nonnormality.

7. Click on the red triangle and select **Normal Quantile Plot** ⇨ **Plot Actual by Quantile**.

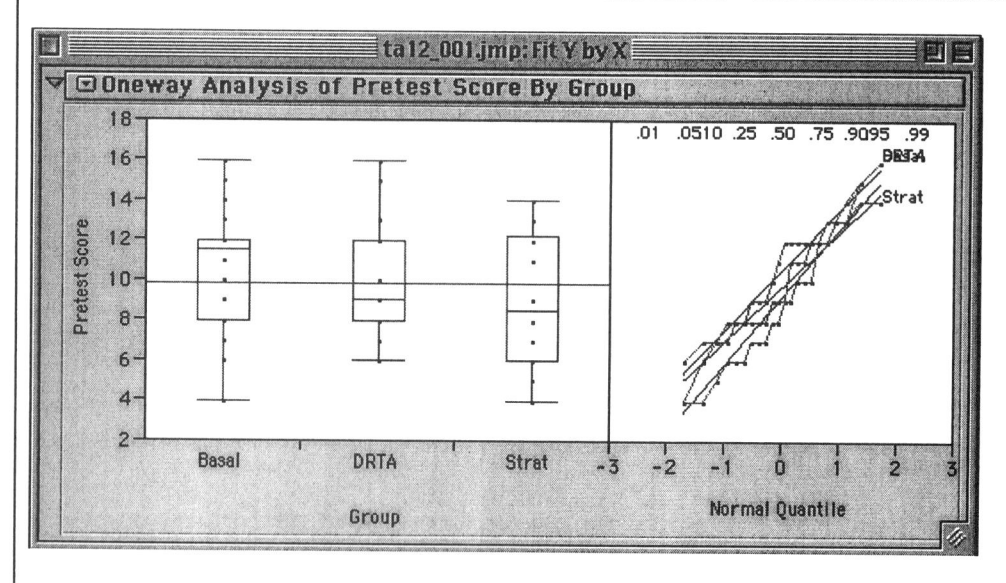

```
┌─────────────────────────────────────────────────────┐
│              ta12_001.jmp: Fit Y by X               │
│ ▽ Oneway Analysis of Pretest Score By Group         │
│                                                     │
│     18                         .01 .05 10 .25 .50 .75 .90 95 .99 │
│     16                                       Basal  │
│     14                                       DRTA   │
│  P  12                                       Strat  │
│  r  10                                               │
│  e   8                                               │
│  t   6                                               │
│  e   4                                               │
│  s   2                                               │
│  t    Basal   DRTA   Strat   -3 -2 -1  0  1  2  3   │
│           Group                 Normal Quantile     │
└─────────────────────────────────────────────────────┘
```

Three normal quantile plots are displayed next to the side-by-side plots, one for each group. The points for our distributions do not deviate from the lines by much and there are no outliers. Because the data look reasonably normal and the standard deviations are about the same, we can compare the three population means using analysis of variance.

The ANOVA and the *F* Test

8. Click on the red triangle and select **Means/Anova/t Test**. (In the display below, the normal quantile plots as well as the boxplots have been deselected.)

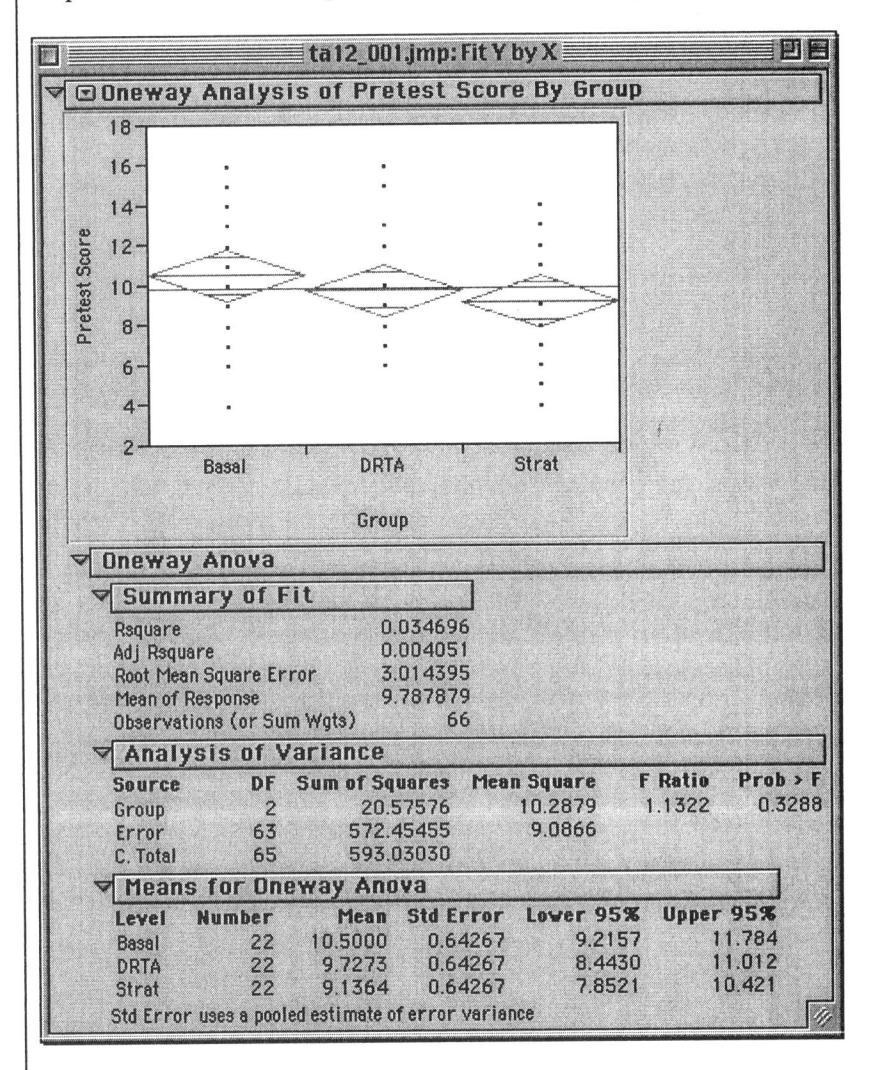

```
ta12_001.jmp: Fit Y by X
Oneway Analysis of Pretest Score By Group
```

[Plot: Pretest Score vs Group (Basal, DRTA, Strat)]

Oneway Anova

Summary of Fit

Rsquare	0.034696
Adj Rsquare	0.004051
Root Mean Square Error	3.014395
Mean of Response	9.787879
Observations (or Sum Wgts)	66

Analysis of Variance

Source	DF	Sum of Squares	Mean Square	F Ratio	Prob > F
Group	2	20.57576	10.2879	1.1322	0.3288
Error	63	572.45455	9.0866		
C. Total	65	593.03030			

Means for Oneway Anova

Level	Number	Mean	Std Error	Lower 95%	Upper 95%
Basal	22	10.5000	0.64267	9.2157	11.784
DRTA	22	9.7273	0.64267	8.4430	11.012
Strat	22	9.1364	0.64267	7.8521	10.421

Std Error uses a pooled estimate of error variance

From the **Analysis of Variance** report, we see that the *F* test statistic (labeled **F Ratio**) is 1.1322 and has 2 and 63 degrees of freedom (**DF**). The *P*-value (labeled **Prob>F**) for the test statistic is 0.3288. Therefore, there is insufficient evidence to conclude that the population means differ (the alternative hypothesis is true). The **Means for Oneway Anova** report lists the three sample means, 10.5, 9.73, and 9.14. The *root mean square error* (3.01) and R^2 (labeled **RSquare** and equal to 3.47%) are found in the **Summary of Fit** report.

Remark

The modeling type of the column that identifies the groups must be "Nominal" (see Section 7.2 in Chapter 7 for details).

12.2 Comparing the Means

The *Analysis of Variance F test* is an overall test to determine if there is good evidence of *any* differences among the means that we wish to compare. If the *F* test is statistically significant, then, without further analysis, all that we can safely conclude is that the groups with the largest and smallest means are significantly different. More detailed follow-up is required to decide which of the other group means differ significantly. *Multiple comparison* methods can be used to compare pairs of population means. In addition, specific questions formulated before examination of the data can be expressed as *contrasts* and inferences made on these *contrasts*. We illustrate this by examining a post-instruction test score from the previous study.

IPS Example 12.10 Comparing instruction methods for reading

The study of reading comprehension in children compared three methods of instruction. After instruction was completed, the students were tested for reading comprehension. There were several post-instruction measures. We analyze the third one here and call it **Comp**; this is short for comprehension.

Let μ_B, μ_D, and μ_S denote the population post-instruction mean **Comp** score for all children who will receive the basal, DRTA, and strategies methods, respectively. Then, we first test the hypothesis

H_o: $\mu_B = \mu_D = \mu_S$ versus
H_a: not all of the μ_i are equal; i.e., at least two of the μ_i differ.

1. Import the data from the file **reading.txt** (in the appendix folder) on the IPS CD-ROM and name the seven columns **Subject, Group, Pre1, Pre2, Post1, Post2**, and **Comp**.

2. Save the data table as **reading.jmp**.

3. Select **Analyze** ⇨ **Fit Y by X**.
   a. Select **Comp** and press **Y, Response**.
   b. Select **Group** and press **X, Factor** and **OK**.

4. Click on the red triangle and select **Means/Anova/t Test**.

The analysis of variance report shows that the *F* statistic = 4.4811 with a *P*-value of 0.0152. We have good evidence to conclude that the three methods of instruction result in different population mean comprehension scores.

Since the basal method has the smallest sample mean ($\bar{x} = 41.1$) and the DRTA method has the largest ($\bar{x} = 46.7$), we can conclude that the DRTA method of instruction results in higher comprehension scores than the basal method. However, what about the strategies method? Is it better than the basal method as well? Also, is the DRTA method better than the strategies method? Multiple comparison procedures answer these questions.

reading.jmp: Fit Y by X

Oneway Analysis of Comp By Group

Oneway Anova

Summary of Fit

Rsquare	0.12454
Adj Rsquare	0.096747
Root Mean Square Error	6.314108
Mean of Response	44.01515
Observations (or Sum Wgts)	66

Analysis of Variance

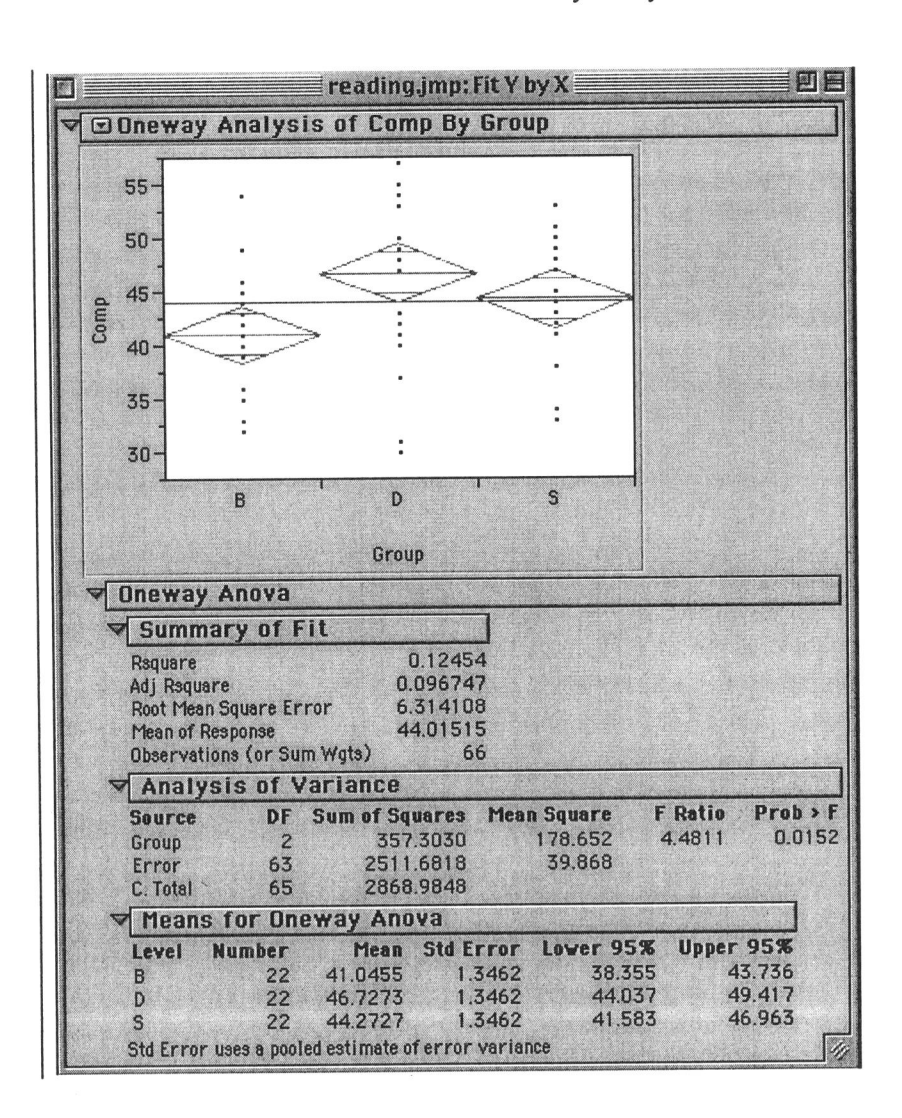

Source	DF	Sum of Squares	Mean Square	F Ratio	Prob > F
Group	2	357.3030	178.652	4.4811	0.0152
Error	63	2511.6818	39.868		
C. Total	65	2868.9848			

Means for Oneway Anova

Level	Number	Mean	Std Error	Lower 95%	Upper 95%
B	22	41.0455	1.3462	38.355	43.736
D	22	46.7273	1.3462	44.037	49.417
S	22	44.2727	1.3462	41.583	46.963

Std Error uses a pooled estimate of error variance

Multiple Comparisons

JMP provides several multiple comparison methods. We illustrate two: the *least-significant differences, or LSD,* method and the *Tukey-Kramer honestly-significant differences, or HSD,* method. The JMP command **Each Pair, Student's t** gives the *least-significant differences* method while the command **All Pairs, Tukey HSD** gives the *Tukey-Kramer honestly-significant differences, or HSD,* method. We recommend using the latter method.

IPS Example 12.6 Which methods differ?

We start with the *least-significant differences*, or repeated two-sample *t* test, procedure.

1. Press the red triangle and select **Compare Means** ⇨ **Each Pair, Student's t**.

Each multiple comparison procedure begins with a *comparison circles* plot, which is a visual representation of group mean comparisons. You can compare each pair of group means visually by clicking on any comparison circle to highlight it. The highlighted circle appears with a thick (red) solid line. Circles representing groups with means that are *not* significantly different from the selected group appear as thin (red) lines. Circles representing means that are significantly different from the selected circle are displayed with a thicker (gray) color. A table of mean comparisons follows.

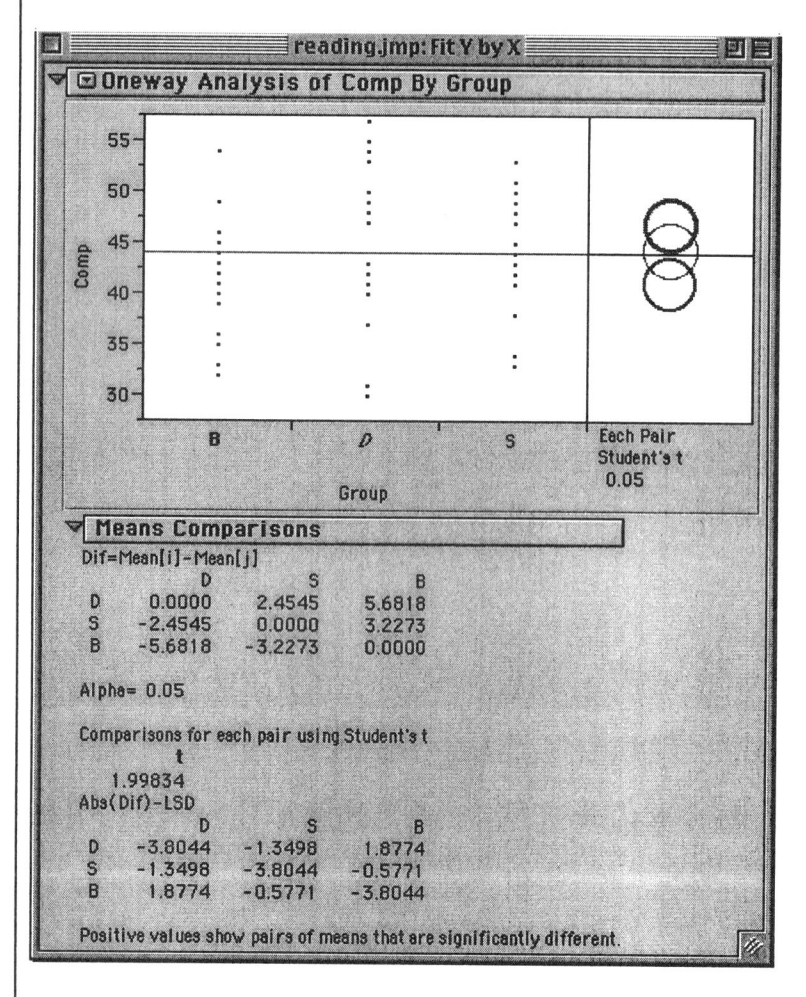

```
                    reading.jmp: Fit Y by X
    Oneway Analysis of Comp By Group

    55-
    50-
    45-
 Comp
    40-
    35-
    30-
           B          D          S       Each Pair
                    Group                Student's t
                                            0.05

    Means Comparisons
    Dif=Mean[i]-Mean[j]
           D         S         B
    D   0.0000    2.4545    5.6818
    S  -2.4545    0.0000    3.2273
    B  -5.6818   -3.2273    0.0000

    Alpha= 0.05

    Comparisons for each pair using Student's t
        t
      1.99834
    Abs(Dif)-LSD
           D         S         B
    D  -3.8044   -1.3498    1.8774
    S  -1.3498   -3.8044   -0.5771
    B   1.8774   -0.5771   -3.8044

    Positive values show pairs of means that are significantly different.
```

2. Click on the *comparison circle* of a group (**Basal** was selected here) to see which group means are significantly different from that group.

The selected circle appears with a thick red line. Since only the circle for DRTA is a thick gray color, only the population mean **Comp** score for the DRTA method differs from that for the basal method.

3. Click on the *comparison circle* of another group, say **Strategies**. The strategies method is not shown to be significantly different from either of the other two methods.

As the textbook states, the method of *least-significant differences, or LSD,* has some undesirable properties. In general, it should not be used when comparing more than two groups because it does not take into consideration all the comparisons that you are making. The *Tukey-Kramer honestly-significant difference, or HSD,* method, unlike the *LSD*, does control the overall probability of some false rejection among all pairs. It is less conservative than the *Bonferroni* method and offers just as much protection. We recommend that it be used in place of the *Bonferroni* method for problems in the textbook. To display the *Tukey-Kramer HSD* procedure,

4. Press the red triangle and select **Compare Means** ⇨ **All Pairs, Tukey HSD**.

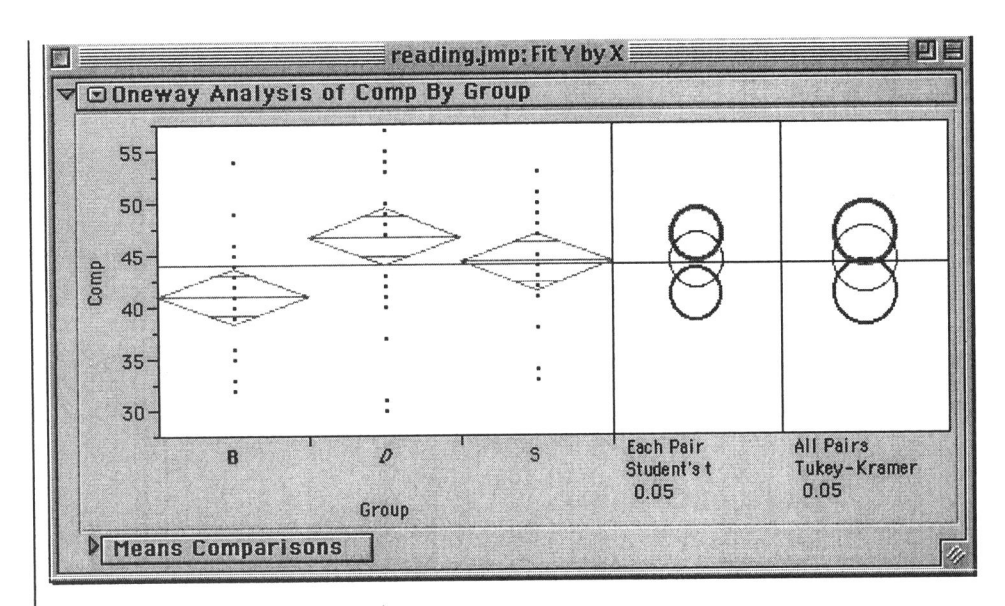

[JMP output window: Oneway Analysis of Comp By Group — dot plot with diamonds for groups B, D, S and means comparison circles for Each Pair Student's t 0.05 and All Pairs Tukey-Kramer 0.05]

For this example, we reach the same conclusions with both methods. When the conclusions differ, the Tukey HSD method is usually the appropriate method.

Remarks

- The default level of significance is 0.05. This can be changed using the **Set Alpha Level** command on the red triangle pop-up menu.

- The abbreviation LSD in the **Means Comparisons** report refers to the smallest difference between two means that is statistically significant; i.e., what the textbook calls *minimum significant difference*, or MSD.

Contrasts

Specific questions formulated before examination of the data can be expressed as *contrasts*. Significance tests provide answers to these questions. We use a new platform in JMP, **Fit Model**, to construct and test *contrasts*.

IPS Example 12.6 Which methods differ? (cont'd.)

The basal method was the standard method commonly used in schools. The DRTA and strategies methods were new methods of teaching that were designed to increase the reading comprehension of the children. The new methods were not identical but each involved the use of similar comprehension techniques. Consequently, prior to analyzing the data, the researchers formulated the following hypotheses to be tested.

(1) H_{o1}: $.5(\mu_D + \mu_S) = \mu_B$ versus the alternative H_{a1}: $.5(\mu_D + \mu_S) > \mu_B$, and
(2) H_{o2}: $\mu_D = \mu_S$ versus the alternative H_{a2}: $\mu_D \neq \mu_S$

Each of these results in a contrast to be tested:

(1) $\psi_1 = -\mu_B + .5\mu_D + .5\mu_S$ and
(2) $\psi_2 = \mu_D - \mu_S$

Let's construct and test these contrasts.

1. Select **Analyze** ⇨ **Fit Model**.
    a. Select **Comp** and press **Y**.
    b. Select **Group** and press **Add** under `Construct Model Effects` and **Run Model**.

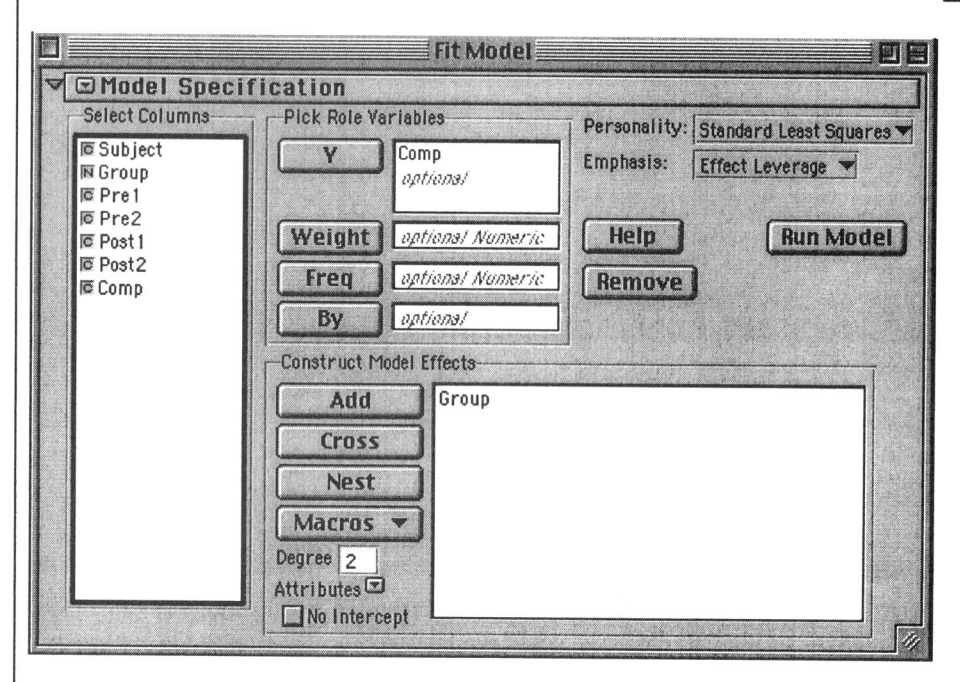

We are interested in the **Group** report in the resulting window. Close the **Whole Model** report and the **Leverage Plot** for **Group** to obtain the following.

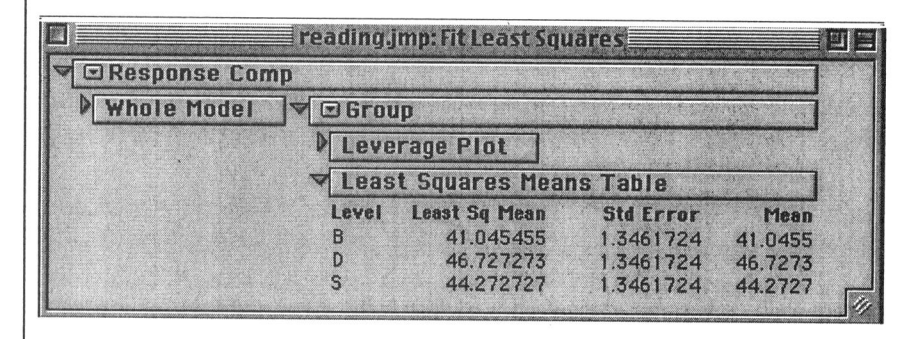

2. Press the red triangle next to the **Group** report title and select **LSMeans Contrast.**

3. To specify the first contrast,
    a. Press the − (minus) sign next to B,
    b. Press the + (plus) sign next to D, and
    c. Press the + (plus) sign next to S.

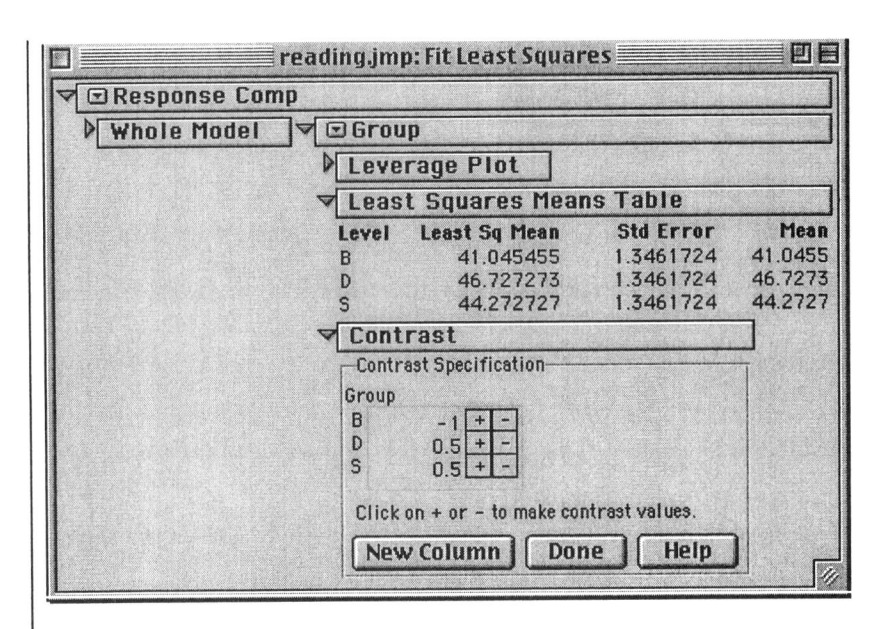

4. Press **Done** and open the resulting **Test Detail** report by clicking the disclosure diamond.

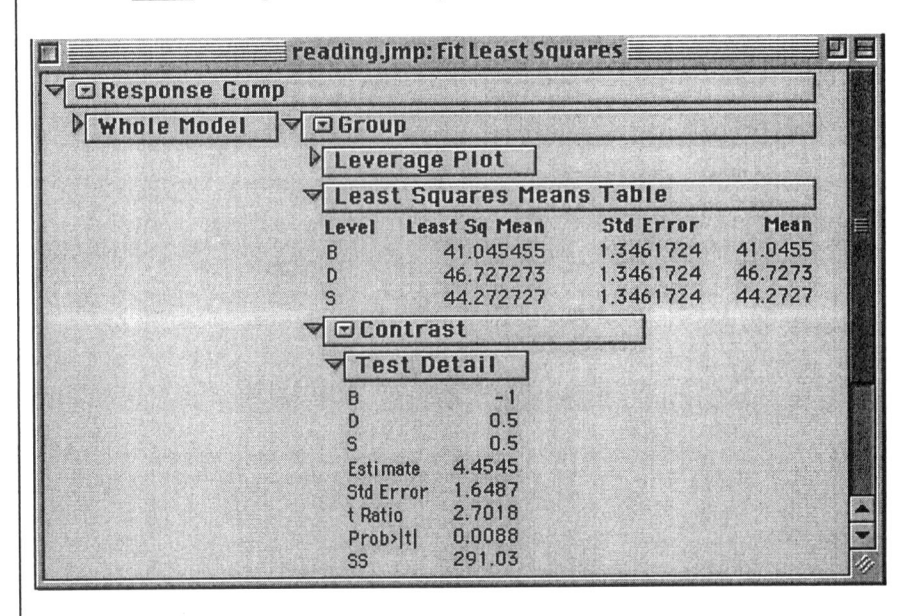

Notice the coefficients of **Basal**, **DRTA**, and **Strategies** are −1, 0.5, and 0.5, respectively; the estimate of the first contrast is 4.45 and its standard error is 1.6487. The *t* statistic is 2.70 and its two-sided *P*-value (labeled **Prob>|t|**) is 0.0088. Since the researchers were interested in showing only that the new methods were better than the standard one, they were interested in a one-sided alternative and the *P*-value for the first hypothesis is 0.0044 (equal to one-half of 0.0088). The data strongly demonstrate that the new methods produce higher mean scores than the old. To obtain the test for the second contrast, simply

5. Select **LSMeans Contrast** from the red triangle menu on the **Group** report title bar again.

6. Press the + (plus) sign next to D and the − (minus) sign next to S.

7. Select **Done** and open the **Test Detail** report.

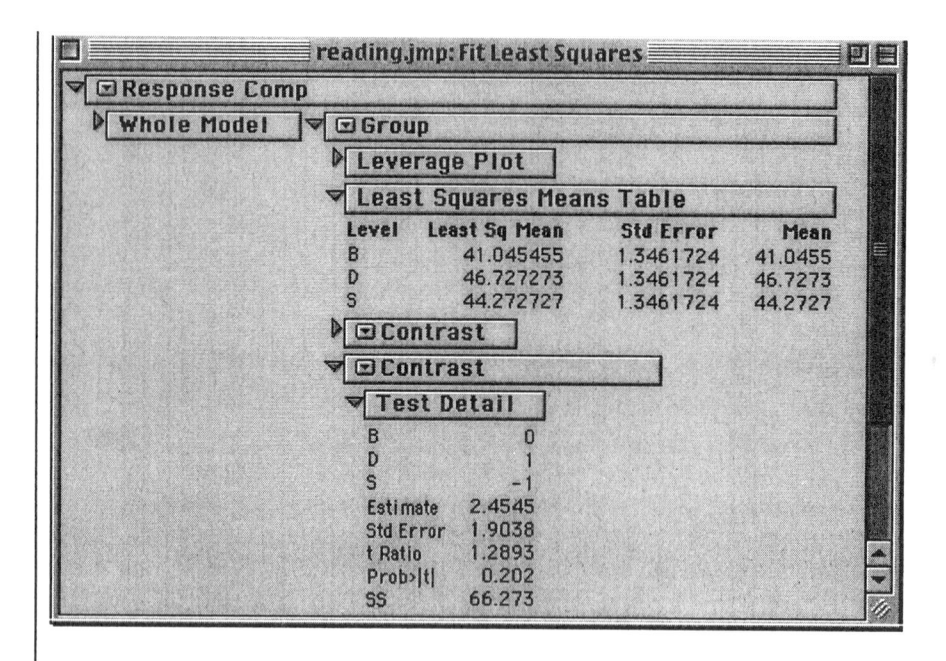

```
┌─────────────────────────────────────────────────┐
│         reading.jmp: Fit Least Squares          │
│ ▽ Response Comp                                 │
│   ▷ Whole Model   ▽ Group                       │
│              ▷ Leverage Plot                    │
│              ▽ Least Squares Means Table        │
│                Level  Least Sq Mean  Std Error   Mean
│                B       41.045455    1.3461724   41.0455
│                D       46.727273    1.3461724   46.7273
│                S       44.272727    1.3461724   44.2727
│              ▷ Contrast
│              ▽ Contrast
│                ▽ Test Detail
│                  B         0
│                  D         1
│                  S        -1
│                  Estimate  2.4545
│                  Std Error 1.9038
│                  t Ratio   1.2893
│                  Prob>|t|  0.202
│                  SS        66.273
└─────────────────────────────────────────────────┘
```

The estimate of the second contrast, which compares the two new methods, is 2.45 with a standard error of 1.90. This gives a *t* statistic equal to 1.2893 and a two-sided *P*-value of 0.202. There is not sufficient evidence to differentiate the effectiveness of the two new methods of instruction.

Remark

The **Fit Model** platform is discussed in more detail in the next chapter.

12.3 Summary

All graphs and statistical computations in this chapter are performed in the **Fit Y by X** platform.

Graph/Computation	Command
ANOVA *F* test	**Fit Y by X ⇨ Means/Anova/t-Test**
Evaluating assumptions	
Normal quantile plots	**Fit Y by X ⇨ Normal Quantile Plots**
Equal variances	**Fit Y by X ⇨ Unequal Variances**
Multiple comparisons	**Fit Y by X ⇨ Compare Means**
Contrasts	**Fit Model ⇨ LSMeans Contrast**

12.4 Exercises

1. The *F* versus *t* Test. In Section 7.2 of Chapter 7, we compared two means using the two-sample *t* test. Let's apply the Analysis of Variance *F* test to the study on the directed reading activity discussed in Example 7.14 and compare the results.

a. Find the value of the *pooled* two-sample t statistic and the associated P-value that compares the mean degree of reading (DRP) test scores for students who take the new directed reading activity and those who do not.
b. Find the value of the ANOVA F statistic and the associated P-value for the same comparison.
c. Compare the answers to parts a and b.
d. Compare the value of the standard error of the difference (labeled **Std Error**) in the **t Test** report with the means square error (in the row labeled **Error** and the column labeled **Mean Square**) in the **Analysis of Variance** report. How are they related?

Use JMP for the following exercises from the textbook.

2. Discounting and price expectation. Exercises 12.16 and 12.17. Let the column containing the number of promotions in the ten-week period have a data type of "Numeric" and change its modeling type to "Nominal." Save the JMP data table for later use.

3. Do piano lessons improve the reasoning in preschool children? Exercises 12.18, 12.19 and 12.20.

4. Breast-feeding and energy intake in infants. Exercises 12.21 and 12.22.

5. Which color do beetles find more attractive? Exercises 12.30, 12.31, 12.59, and 12.60.

6. The effectiveness of dandruff shampoos. Exercises 12.48 through 12.50.

7. Transforming data. Exercises 12.53 and 12.54.

8. Improving reading comprehension. Exercises 12.55 through 12.58.

9. Predicting price expectations. Exercise 12.60. Change the modeling type of the column that contains the number of promotions in the ten-week period to "Continuous."

Chapter 13

Two-Way Analysis of Variance

In Chapter 12, inference for the relationship between a response variable *y* and a *single factor, or categorical explanatory* variable *x,* was discussed. Now, we consider the case when there is more than one factor or explanatory variable—specifically, when there are *two factors*.

13.1 The Two-Way ANOVA Model

While the plots and calculations in this section can easily be carried out using JMP, I suggest that you use paper, pencil, and a simple calculator at first. Few calculations are involved and you will gain a better feel for the new concepts introduced in this section—specifically, main effects and interaction.

13.2 Inference for Two-Way ANOVA

Since there is more than one factor, or explanatory variable, we use the JMP **Fit Model** platform to analyze the data in this chapter.

IPS Example 13.8 Lifestyles and cardiac fitness

A study of cardiovascular risk factors compared runners who averaged at least 15 miles per week with a control group described as 'generally sedentary'. Both men and women were included in the study. One variable of interest in the study was the heart rate after 6 minutes of exercise on a treadmill.

1. Import the text file **eg13_008.txt** on the IPS CD-ROM and name the variables **Subject, Group, Gender,** and **Heart Rate**.

2. Save the JMP data table as **eg13_008.jmp**. There should be 800 individuals and 4 variables.

Use the **Distribution** platform in JMP to first look at the distributions of the three variables.

3. Select **Analyze** ⇨ **Distribution**.

4. Select **Group**, **Gender**, and **Heart Rate**; press **Y, Columns** and **OK**.

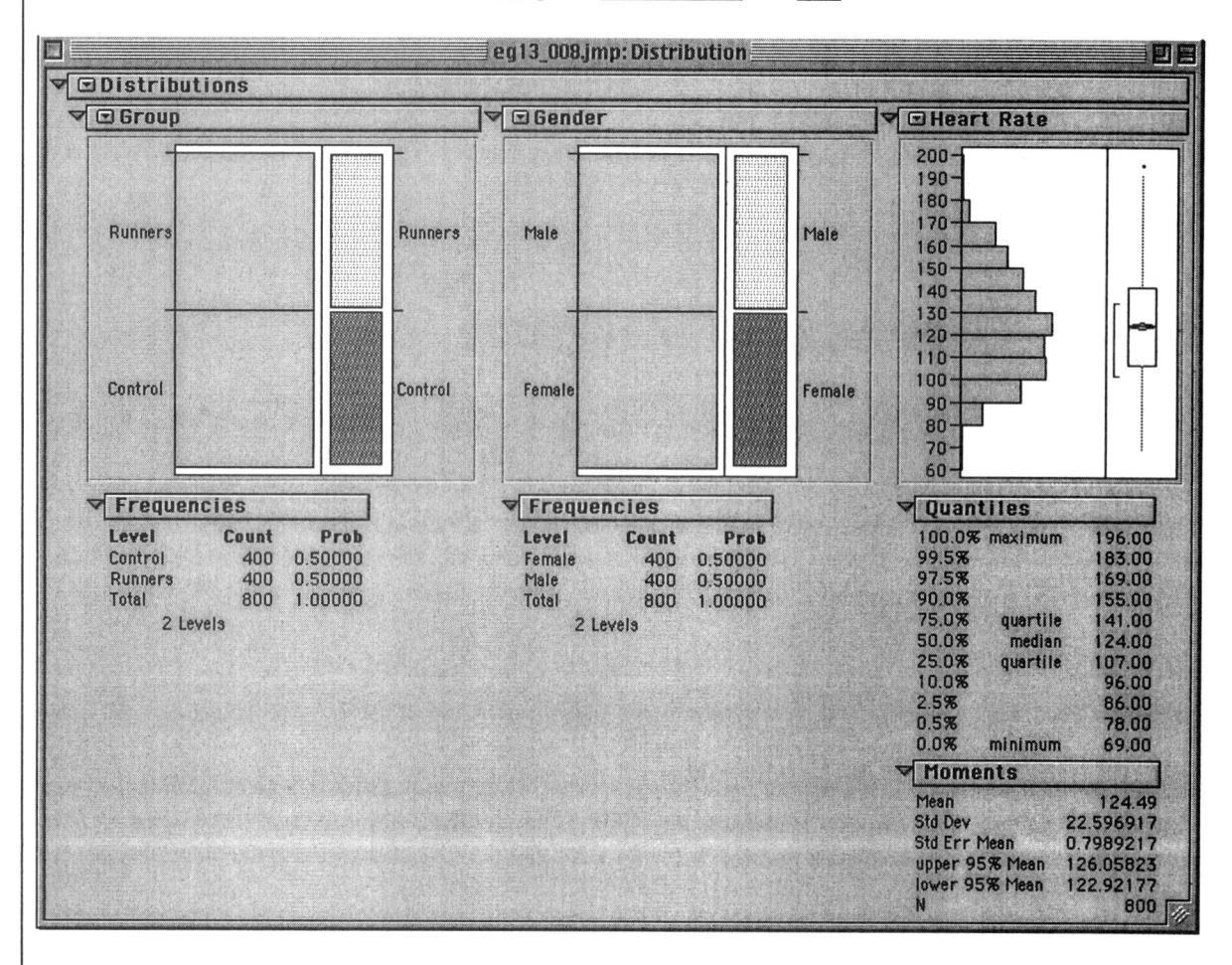

This was a carefully designed study with an equal number of males and females and an equal number of runners and nonrunners. Further examination (**Fit Y by X** with **Group** and **Gender**) would show that there are exactly 200 people in each of the four group-by-gender categories. The response variable **Heart Rate** is symmetrically distributed with a mean of 124.49 beats per minute (bpm) and a standard deviation of 22.60 bpm. Subject 154 is somewhat outlying.

It is possible to examine the relationships of **Heart Rate** to **Group** and to **Gender** separately (using **Fit Y by X**); however, this fails to detect a possible interaction of **Group** and **Gender**. We proceed directly to a model that includes both **Gender** and **Group** as factors.

5. Select **Analyze** ⇨ **Fit Model**.
   a. Select **Heart Rate** and press **Y**.
   b. Select **Group** and **Gender** and press **Add**.
   c. Select **Group** and **Gender** and press **Cross** to include an interaction effect.
   d. Press **Run Model**.

Examine the residuals. First, scroll down to the **Residual by Predicted Plot**. The spread of the four groups appears to be roughly the same. To check the residuals for normality, save them to the data table and use the **Normal Quantile Plot** command.

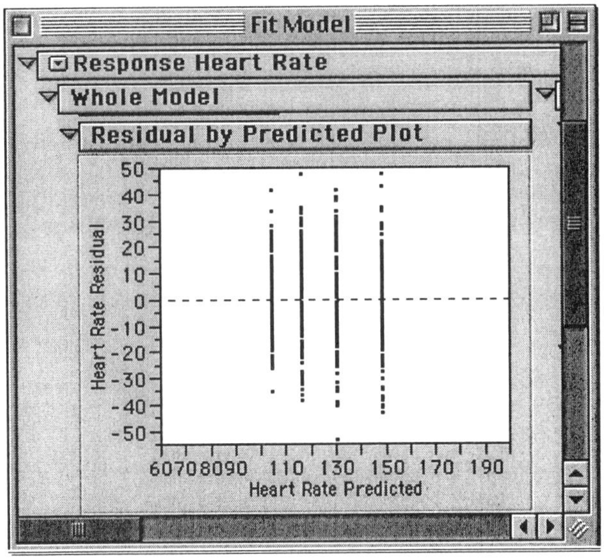
Fit Model
Response Heart Rate
Whole Model
Residual by Predicted Plot
Heart Rate Residual
50
40
30
20
10
0
-10
-20
-30
-40
-50
60708090 110 130 150 170 190
Heart Rate Predicted

6. Press the red triangle next to **Response Heart Rate** and select **Save Columns** ⇨ **Residuals**.

7. Select **Analyze** ⇨ **Distribution** ⇨ **Residuals Heart Rate** and press **Y, Columns** and **OK**.

8. Select **Normal Quantile Plot** from the red triangle menu.

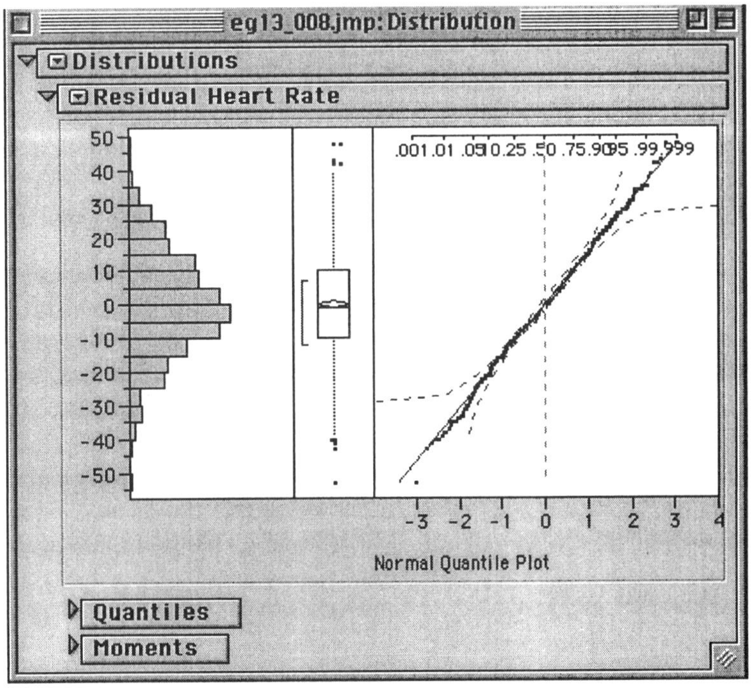
eg13_008.jmp: Distribution
Distributions
Residual Heart Rate
50
40
30
20
10
0
-10
-20
-30
-40
-50
.001 .01 .05 .10 .25 .50 .75 .90 .95 .99 .999
-3 -2 -1 0 1 2 3 4
Normal Quantile Plot
Quantiles
Moments

The data appear to be reasonably normal. Now inspect the **Whole Model** report.

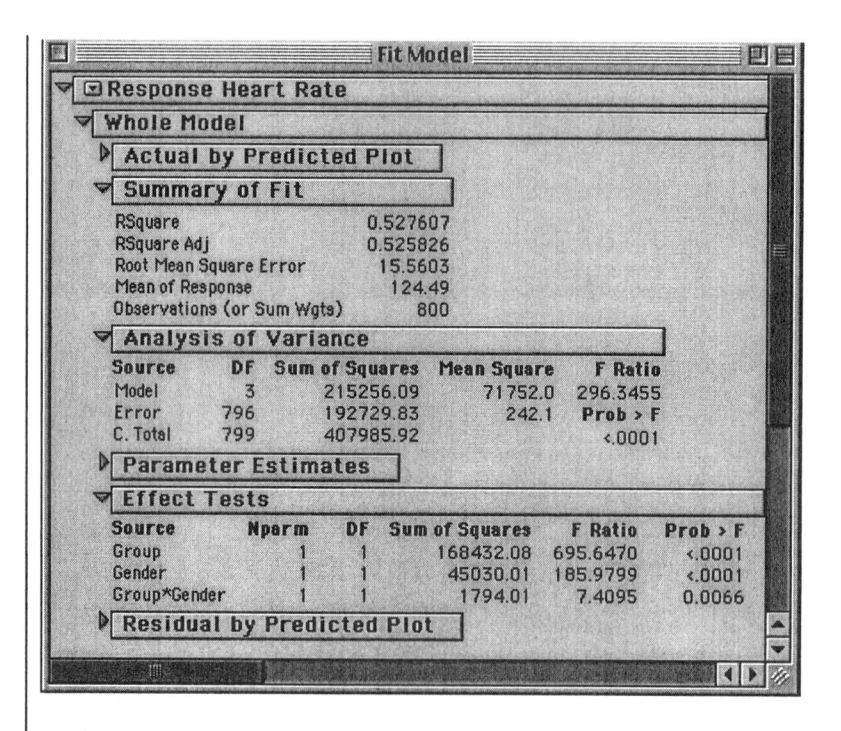

```
         ┌──────────────── Fit Model ─────────────────┐
         │ ▽ Response Heart Rate                       │
         │   ▽ Whole Model                             │
         │     ▷ Actual by Predicted Plot              │
         │     ▽ Summary of Fit                        │
         │        RSquare                    0.527607  │
         │        RSquare Adj                0.525826  │
         │        Root Mean Square Error     15.5603   │
         │        Mean of Response           124.49    │
         │        Observations (or Sum Wgts) 800       │
         │     ▽ Analysis of Variance                  │
         │       Source  DF  Sum of Squares  Mean Square  F Ratio    │
         │       Model    3     215256.09      71752.0   296.3455   │
         │       Error  796     192729.83        242.1   Prob > F   │
         │       C.Total 799    407985.92                 <.0001    │
         │     ▷ Parameter Estimates                   │
         │     ▽ Effect Tests                          │
         │       Source       Nparm  DF  Sum of Squares  F Ratio  Prob > F │
         │       Group          1     1    168432.08    695.6470   <.0001  │
         │       Gender         1     1     45030.01    185.9799   <.0001  │
         │       Group*Gender   1     1      1794.01      7.4095   0.0066  │
         │     ▷ Residual by Predicted Plot            │
         └─────────────────────────────────────────────┘
```

In the **Summary of Fit** report, the overall mean heart rate is 124.5 bpm and the root mean square error is 15.6 bpm. The *coefficient of determination*, R^2, equals .528. Thus, **Gender** and exercise **Group** explain about 53% of the variability in the heart rates.

The **Analysis of Variance** report is, in effect, a one-way ANOVA table with four treatments: female control, female runner, male control, and male runner. The F test statistic (296.3) and its associated *P*-value (< .0001) indicate that the population means of the four treatments are not the same.

The **Effect Tests** report contains the sum of squares, degrees of freedom, and F statistics for the **Group** and **Gender** main effects and the **Group-by-Gender** interaction. These sum of squares and degrees of freedom partition the sum of squares and degrees of freedom for Model. Examine the *P*-values (labeled as **Prob > F**). The interaction is strongly significant. To interpret the results, we examine a *profile plot* of the means.

9. Click on the red triangle next to the **Group*Gender** effect report and select **LSMeans Plot.**

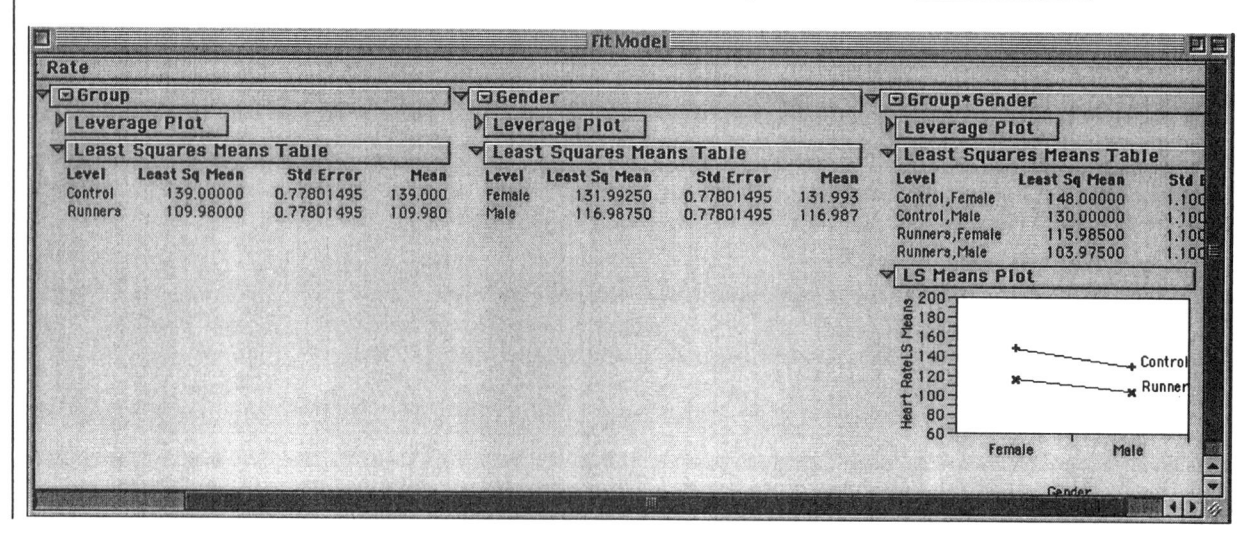

```
┌─────────────────────────────── Fit Model ────────────────────────────────┐
│ Rate                                                                      │
│ ▽ Group                  │ ▽ Gender                │ ▽ Group*Gender       │
│   ▷ Leverage Plot        │   ▷ Leverage Plot       │   ▷ Leverage Plot    │
│   ▽ Least Squares Means Table │ ▽ Least Squares Means Table │ ▽ Least Squares Means Table │
│   Level   Least Sq Mean  Std Error   Mean │ Level  Least Sq Mean  Std Error  Mean │ Level          Least Sq Mean    Std E │
│   Control   139.00000   0.77801495  139.000 │ Female 131.99250   0.77801495 131.993 │ Control,Female    148.00000    1.100 │
│   Runners   109.98000   0.77801495  109.980 │ Male   116.98750   0.77801495 116.987 │ Control,Male      130.00000    1.100 │
│                                             │                                       │ Runners,Female    115.98500    1.100 │
│                                             │                                       │ Runners,Male      103.97500    1.100 │
│                                             │                                       │ ▽ LS Means Plot                      │
│                                             │                                       │    [profile plot: Control and Runner │
│                                             │                                       │     lines by Female/Male]            │
└──────────────────────────────────────────────────────────────────────────┘
```

The two lines in the plot are not parallel. There is interaction. The difference between controls and runners in mean heart rates is greater for women than for men. The interaction, while significant, is not large. Return to the **Effect Tests** report and notice that the main effects for group and gender are both very significant. Runners, on average, have lower heart rates than the controls for both males and females by about 29 bpm (= 139 − 109.98). Women, on average, have higher heart rates than men do by about 15 bpm.

Remark

JMP provides a shortcut for adding the two main effects and the interaction effect to the model in the **Fit Model** dialog. For the previous example, simply select **Group** and **Gender** and press **Macros** ⇨ **Full Factorial**.

13.3 Summary

All graphs and statistical computations for multiple regression models use the **Fit Model** platform. The residuals can be examined using the **Distribution** platform.

Activity	Command
Inference about the model	**Analyze** ⇨ **Fit Model**
Saving residuals	**Fit Model** ⇨ **Save Columns** ⇨ **Residuals**
Profile plots	**Fit Model** ⇨ **LSMeans Plot**

13.4 Exercises

1. Paper and pencil. Exercises 13.7 through 13.12.

2. Repairing serious wounds. Exercises 13.13 through 13.15.

3. Iron-deficiency anemia. Exercises 13.16 and 13.18.

4. Discounting and price expectations. Exercises 13.21 and 13.22.

5. The effect of water and species on nitrogen content. Exercises 13.27 through 13.30.

Chapter 14

Nonparametric Tests

All methods for inference about the means of quantitative response variables discussed thus far have assumed that the variables have normal distributions in the populations or populations from which they are drawn. Nonparametric methods can be used when this assumption fails. The introductory version of JMP, JMP INTRO, has only the *Wilcoxon signed rank test*. JMP IN and the professional version of JMP provide a wide range of nonparametric tests.

14.1 The Wilcoxon Rank Sum Test

The *Wilcoxon rank sum test* compares two distributions to assess whether one has systematically larger values than the other. Only JMP IN and the professional version of JMP provide this nonparametric test. Use the **Nonparametric ⇨ Wilcoxon Test** command on the **Fit Y by X** platform to display the *Wilcoxon rank sum test*. For IPS Example 14.1, the resulting report window is

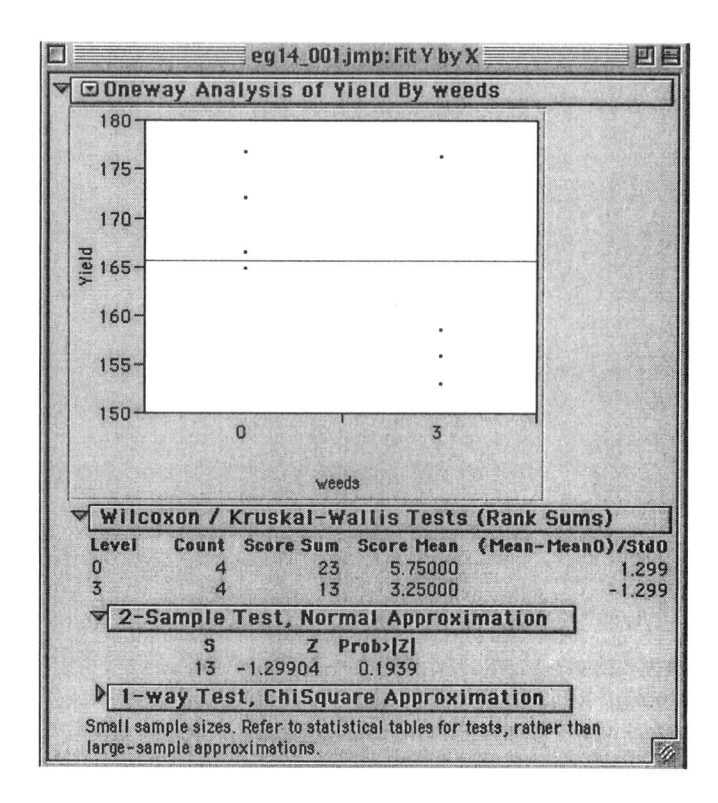

[JMP output screenshot: Oneway Analysis of Yield By weeds, Wilcoxon / Kruskal-Wallis Tests (Rank Sums), 2-Sample Test Normal Approximation showing S=13, Z=-1.29904, Prob>|Z|=0.1939]

14.2 The Wilcoxon Signed Rank Test

The nonparametric *Wilcoxon signed rank test* is often used in place of the *t* test when the assumption of normality does not hold in the one-sample or matched pairs settings. The *Wilcoxon signed rank test statistic* and associated *P*-values can be requested using the **Test Mean** command in the **Distribution** platform.

IPS Example 14.11 Tournament golf scores

Golf scores of 12 members of a college women's golf team for two rounds of tournament play were collected to determine whether player scores in the second round were systematically different (lower or higher) from the first round. The hypothesis to be tested is that in a large population of collegiate woman golfers

H_o: scores have the same distribution in rounds 1 and 2 versus
H_a: scores are systematically higher or lower in round 2.

This is a matched pairs design (see Section 7.1 in Chapter 7). To compare the rounds of golf, we create a column of the differences between the scores for the two rounds of golf and test whether the mean difference is zero.

1. Create a new JMP data table with the variables **Player**, **Round 1**, and **Round 2**.

2. Enter the values for these three variables as follows.

	Player	Round 1 Score	Round 2 Score	Difference
1	1	89	94	5
2	2	90	85	-5
3	3	87	89	2
4	4	95	89	-6
5	5	86	81	-5
6	6	81	76	-5
7	7	102	107	5
8	8	105	89	-16
9	9	83	87	4
10	10	88	91	3
11	11	91	88	-3
12	12	79	80	1

3. Select **Cols** ⇨ **New Column**.
   a. Enter **Difference** for the column name.
   b. Press **New Property** ⇨ **Formula**.
   c. Select **Round 2** from the list of columns, press – (minus) on the keypad, and then select **Round 1** from the list of columns.
   d. Press **OK** and **OK**.

4. Select **Analyze** ⇨ **Distribution** ⇨ **Difference**; press **Y, Columns** and **OK**.

To test the alternative hypothesis that the mean difference is not zero,

5. Select **Test Mean** from the red triangle menu on the **Difference** report title bar.

6. Check the box for the **Wilcoxon Sign-Rank nonparametric test** and press **OK**.

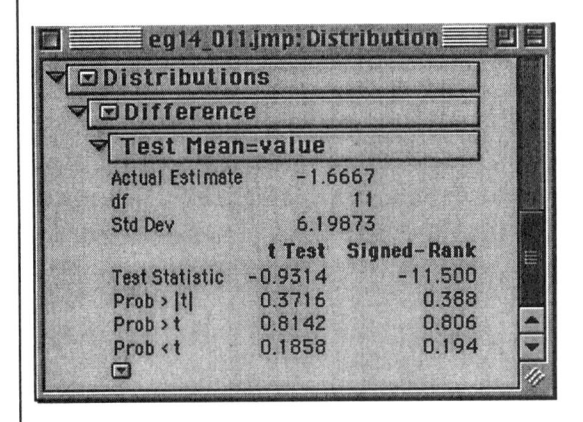

```
Test Mean=value
Actual Estimate   -1.6667
df                     11
Std Dev           6.19873
                t Test    Signed-Rank
Test Statistic  -0.9314       -11.500
Prob > |t|       0.3716         0.388
Prob > t         0.8142         0.806
Prob < t         0.1858         0.194
```

The value of the *Wilcoxon signed-rank test statistic* is –11.500 (JMP gives the deviation from the expected value of 39) and the *P*-value for a two-sided test is 0.388. There is no evidence of a systematic difference between the first and second round golf scores.

14.3 The Kruskal-Wallis Test

The *Kruskal-Wallis test* compares the distributions of several populations based on independent random samples. Only JMP IN and the professional version of JMP provide this nonparametric test. Use the **Nonparametric ⇨ Wilcoxon Test** command on the **Fit Y by X** platform to display the *Kruskal-Wallis test*. For IPS Example 14.13, the resulting report window is

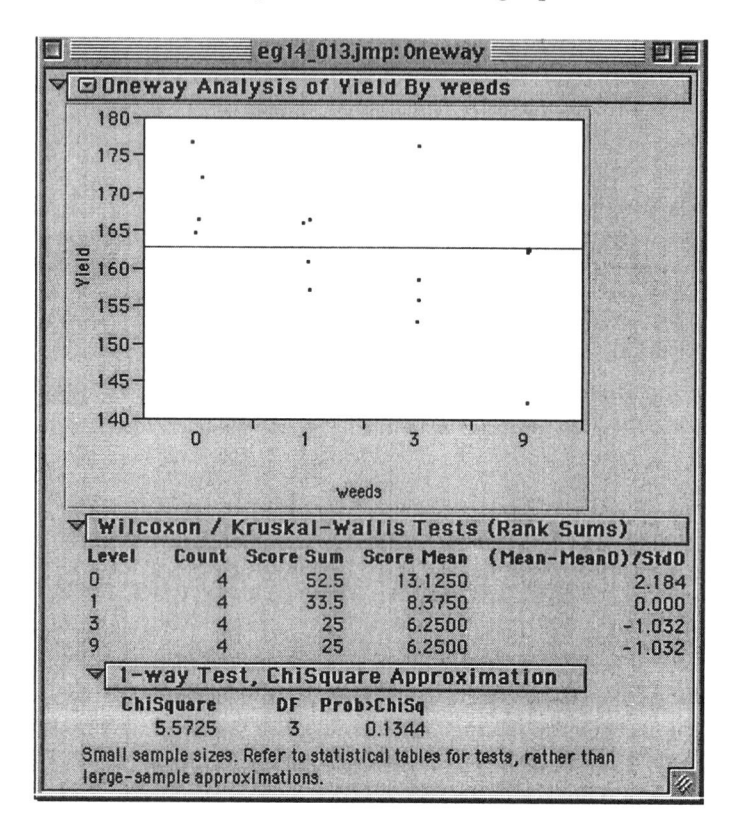

The *Kruskal-Wallis* statistic (labeled **ChiSquare**) equals 5.5725 and the *P*-value equals 0.1344; this is the same as that given in the textbook.

14.4 Exercises

1. Stepping up your heart rate. Exercises 14.13 and 14.14.

2. Attitudes toward food safety at outdoor fairs and festivals. Exercises 14.17 and 14.18.

3. Does direction matter? Exercise 14.21.

Chapter 15

Logistic Regression

The linear regression methods studied in Chapters 10 and 11 modeled the relationship between a *quantitative* response variable and one or more explanatory variables. This chapter discusses similar methods for use when the response variable is *binary*; i.e., has only two possible values.

15.1 The Logistic Regression Model

In this section of the IPS textbook, the *logistic regression* model, *odds*, *odds ratio,* and methods for fitting the model when the explanatory variable is *binary* are discussed.

15.2 Inference for Logistic Regression

Methods of inference for the regression coefficients are discussed in this section. We will use the **Fit Model** platform in JMP to fit and analyze logistic regression models.

IPS Examples 15.1 and 15.6 Binge drinking and gender

Are men and women college students equally likely to be frequent binge drinkers? A survey of over 17,000 students in U.S. four-year colleges collected information on drinking behavior. This data was discussed earlier in Example 8.9.

1. Open the JMP data table **eg08_009.jmp** that was saved in Chapter 8.

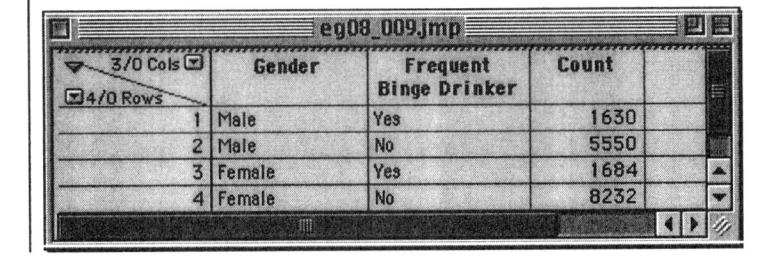

	Gender	Frequent Binge Drinker	Count
1	Male	Yes	1630
2	Male	No	5550
3	Female	Yes	1684
4	Female	No	8232

Let's look at the distribution of the variables **Gender** and **Frequent Binge Drinker**.

2. Select **Analyze** ⇨ **Distribution**.
    a. Select **Gender** and **Frequent Binge Drinker** and press **Y, Columns**.
    b. Select **Count** and press **Freq**.
    c. Press **OK**.

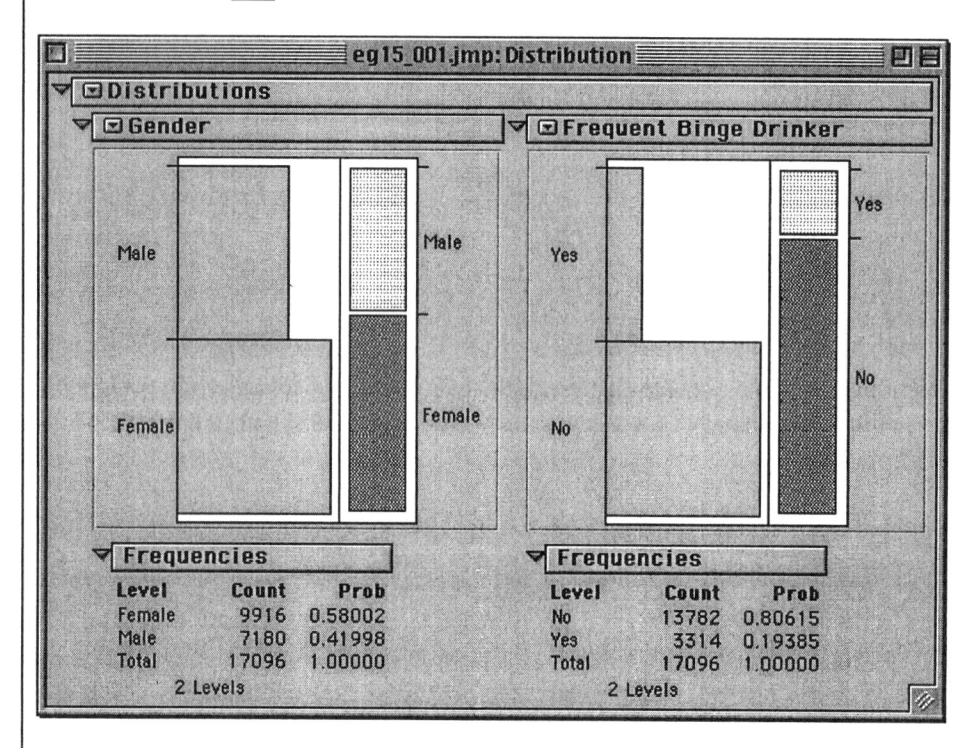

About 19% of the students surveyed were frequent binge drinkers and about 58% of the students were female. To fit a logistic regression model to the data, simply

3. Select **Analyze** ⇨ **Fit Model**.

4. Select **Frequent Binge Drinker** from the list of columns and press **Y**.

5. Select **Gender** and press **Add**.

6. Select **Count** and press **Freq**.

7. Press **Run Model**.

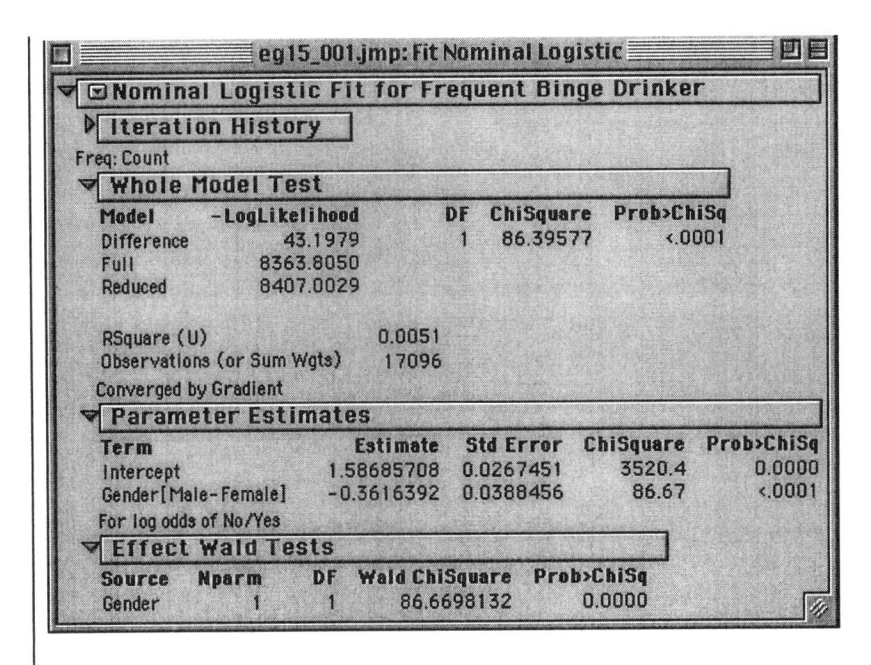

```
┌─────────────eg15_001.jmp: Fit Nominal Logistic──────────┐
│ ▽ Nominal Logistic Fit for Frequent Binge Drinker       │
│   ▷ Iteration History                                   │
│   Freq: Count                                           │
│   ▽ Whole Model Test                                    │
│     Model      -LogLikelihood    DF  ChiSquare  Prob>ChiSq │
│     Difference    43.1979         1   86.39577   <.0001 │
│     Full        8363.8050                               │
│     Reduced     8407.0029                               │
│                                                         │
│     RSquare (U)                  0.0051                 │
│     Observations (or Sum Wgts)   17096                  │
│   Converged by Gradient                                 │
│   ▽ Parameter Estimates                                 │
│     Term               Estimate   Std Error  ChiSquare  Prob>ChiSq │
│     Intercept         1.58685708  0.0267451   3520.4    0.0000  │
│     Gender[Male-Female] -0.3616392 0.0388456   86.67    <.0001  │
│     For log odds of No/Yes                              │
│   ▽ Effect Wald Tests                                   │
│     Source   Nparm  DF  Wald ChiSquare  Prob>ChiSq      │
│     Gender     1     1    86.6698132     0.0000         │
└─────────────────────────────────────────────────────────┘
```

The parameter estimates are given as $b_0 = 1.5869$ and $b_1 = -0.3616$. These have signs opposite those given in the book since JMP has chosen the first value of Gender, "Female," as success. The standard errors are 0.267 and 0.0388. We test the hypothesis

$$H_o: \beta_1 = 0 \quad \text{versus} \quad H_a: \beta_1 \neq 0,$$

with the *Wald ChiSquare statistic* $X^2 = 86.67$ with a *P*-value of <0.001. There is strong evidence that different percentages of men and women are frequent binge drinkers. To obtain the *odds ratio* and 95% confidence intervals for β_1 and it,

8. Press the red triangle, select **Confidence Intervals** ⇨ **0.05**, and press **OK**.

9. Press the red triangle again and select **Odds Ratios**.

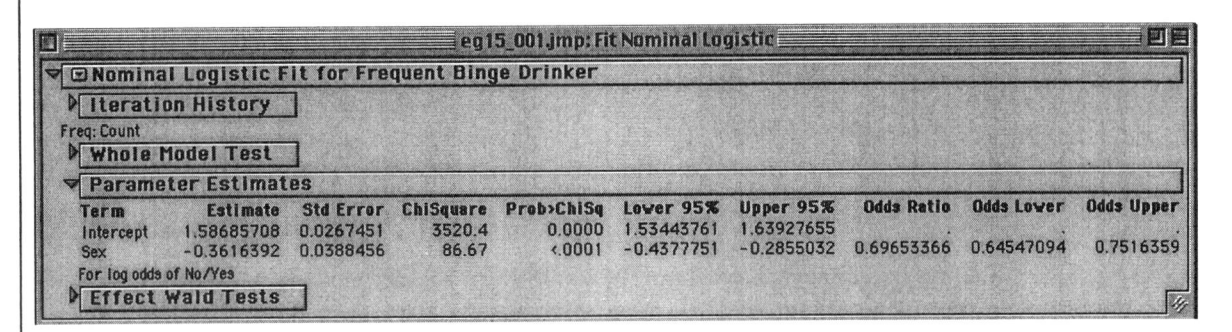

```
┌─────────────eg15_001.jmp: Fit Nominal Logistic──────────────────────────┐
│ ▽ Nominal Logistic Fit for Frequent Binge Drinker                       │
│   ▷ Iteration History                                                   │
│   Freq: Count                                                           │
│   ▷ Whole Model Test                                                    │
│   ▽ Parameter Estimates                                                 │
│    Term      Estimate    Std Error  ChiSquare  Prob>ChiSq  Lower 95%   Upper 95%   Odds Ratio  Odds Lower  Odds Upper │
│    Intercept 1.58685708  0.0267451   3520.4    0.0000     1.53443761  1.63927655                              │
│    Sex      -0.3616392   0.0388456   86.67    <.0001     -0.4377751  -0.2855032   0.69653366  0.64547094  0.7516359 │
│    For log odds of No/Yes                                               │
│   ▷ Effect Wald Tests                                                   │
└─────────────────────────────────────────────────────────────────────────┘
```

A 95% confidence interval for the slope (log odds ratio) is (−0.4378, −0.2855). The odds ratio of women to men is 0.697 and a 95% confidence interval for it is (0.645, 0.752). The odds ratio of men to women and its 95% confidence interval is the reciprocal of these: 1.436 and (1.33, 1.55).

Remark

The **Fit Y by X** platform can be used for fitting and inference of logistic regression models for a *single* explanatory variable. However, it does not provide estimates of the odds ratios and the explanatory variable must have a modeling type of "Continuous." It does provide a plot of the probability of success versus the explanatory variable.

15.3 Multiple Logistic Regression

In multiple logistic regression, the response variable has only two possible values and there can be several explanatory variables. The **Fit Model** platform is used for analysis and the *Wald statistic* plays a role similar to that of the *t statistic* in multiple linear regression.

IPS Example 15.10 What makes cheddar cheese tasty?

As cheddar cheese matures, a variety of chemical processes take place. The taste of matured cheese is related to the concentration of several chemicals in the final product. In an Australian study, samples of cheddar cheese were analyzed for their chemical composition and were subjected to taste tests. For this example, the cheese is classified as acceptable if **Taste** $\geq$ 37 and unacceptable otherwise. We wish to predict the odds that the cheese is acceptable using the three variables **Acetic**, **H2S**, and **Lactic**.

1. Open the text file **Cheese.txt** on the IPS CD-ROM and name the variables **Case**, **Taste**, **Acetic**, **H2S**, and **Lactic**.

Create the variable **TasteOk** with the values 1 and 0. Since the **Fit Model** platform fits the log odds of the smallest numeric value of the response variable (zero in this case), we assign **TasteOk** = 0 when **Taste** $\geq$ 37. You can enter the values directly or use the Formula Editor (see Section 0.3.3 in Chapter 0 and below).

2. Select **Cols** $\Rightarrow$ **New Column**.
   a. Enter `TasteOk` in the `Column Name` field.
   b. Press **New Property** $\Rightarrow$ **Formula**.
   c. Enter the following formula and press **OK**.

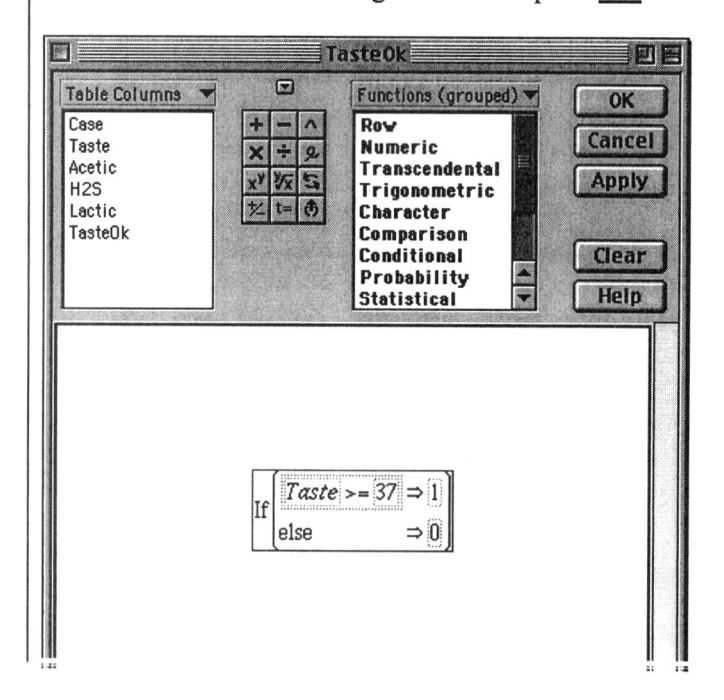

To fit the model, we use the **Fit Model** platform and include all the explanatory variables as `Model Effects`.

3. Select **Analyze** ⇨ **Fit Model**.
   a. Select **TasteOk** from the list of columns and press **Y**.
   b. Select **Acetic**, **H2S**, and **Lactic**, and press **Add**.
   c. Press **Run Model**.

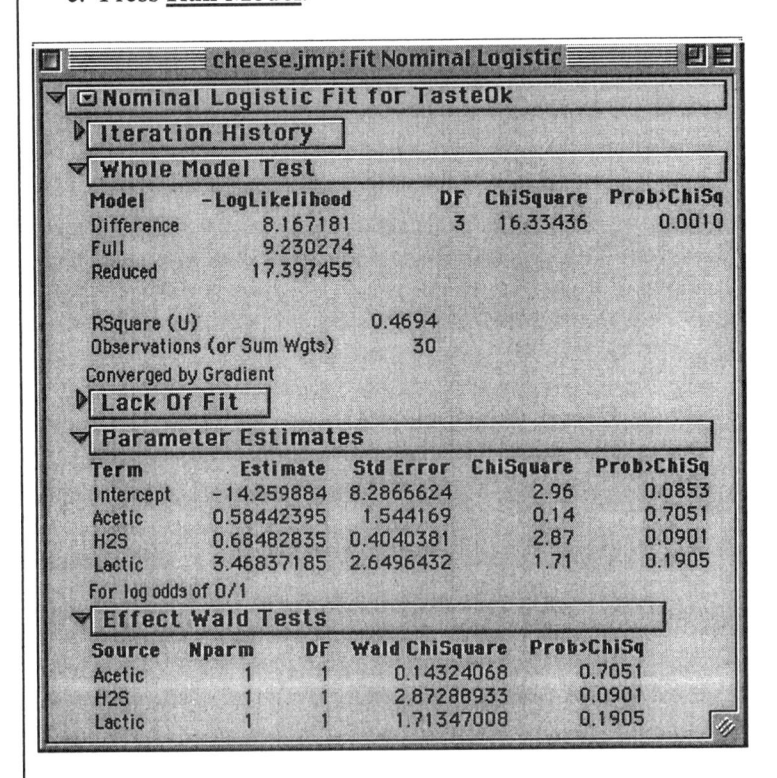

```
cheese.jmp: Fit Nominal Logistic
Nominal Logistic Fit for TasteOk
▶ Iteration History
▼ Whole Model Test
Model       -LogLikelihood    DF   ChiSquare   Prob>ChiSq
Difference      8.167181       3    16.33436     0.0010
Full            9.230274
Reduced        17.397455

RSquare (U)                  0.4694
Observations (or Sum Wgts)       30
Converged by Gradient
▶ Lack Of Fit
▼ Parameter Estimates
Term        Estimate     Std Error    ChiSquare   Prob>ChiSq
Intercept  -14.259884    8.2866624      2.96        0.0853
Acetic       0.58442395  1.544169       0.14        0.7051
H2S          0.68482835  0.4040381      2.87        0.0901
Lactic       3.46837185  2.6496432      1.71        0.1905
For log odds of 0/1
▼ Effect Wald Tests
Source   Nparm   DF   Wald ChiSquare   Prob>ChiSq
Acetic     1      1      0.14324068      0.7051
H2S        1      1      2.87288933      0.0901
Lactic     1      1      1.71347008      0.1905
```

From the **Parameter Estimates** report, we see that the fitted model is

log (Odds of acceptable cheese) = −14.26 + 0.58 Acetic + 0.68 H2S + 3.47 Lactic.

Examine the **Whole Model Test** report. The hypothesis that all the logistic regression coefficients are zero is tested by the loglikelihood test statistic (labeled **ChiSquare** and equal to 16.33). The *P*-value (**Prob>ChiSq**) is 0.0010. We can thus conclude that at least one of the chemicals can be used to predict the odds that the cheese is acceptable.

Examine the coefficients of each variable and the tests that each is zero when the other variables are in the model. The test statistics in the column **Wald ChiSquare** of the **Effect Wald Tests** report are the statistics given in the textbook. The *P*-values are 0.7051, 0.0901, and 0.1905. If two of the three variables are in the model, the third is of little to no value.

15.4 Summary

All graphs and statistical computations for multiple regression models use the **Fit Model** platform.

Activity	Command
Inference about the model	**Analyze ⇨ Fit Model**
Confidence intervals	**Fit Model ⇨ Confidence Intervals**
Odds ratio	**Fit Model ⇨ Odds Ratios**

15.5 Exercises

You should use JMP on any problem in the IPS text for which the data are provided. Here are some suggestions:

1. Insecticide dosage. Examples 15.7 and 15.8.

2. Is gender related to alcohol use in bicycle fatalities? Exercise 15.17.

3. What makes cheddar cheese tasty? Exercises 15.18 through 15.20.

4. Predicting a high GPA. Exercises 15.21 through 15.24.

5. Safe hospitals. Exercise 15.25.